The Synthetic Biosphere

From Petrochemical Invasion to Engineered Molecular Erasure

By Dr. Damon Rahimi Fard

Contents

The Synthetic Encrustation: Navigating the Era of Universal Polymer Ubiquity 1

Mapping the Global Pipeline: From Land to Sea 3

Mechanical Segregation: The Cascade of Exclusion 4

Electronic and Molecular Sorting: The High-Tech Frontier......................5

Gravimetric Defiance: Navigating the Buoyancy Threshold 7

The Final Convergence: A Unified Theory of Detection............................7

The Forensic Identification Crisis: Decoding the Synthetic Signature.........8

The Synthetic Trophic Shift: Dynamics of Polymer Integration and Marine Stratification...10

The Synthetic Canopy: Thermodynamic and Biogeochemical Disruption of the Oceanic Engine 13

The Biological Trojan: Synthetic Bio-Accumulation and the Erasure of Species Boundaries 16

The Polymer Terminal: Geopolitical Inertia and the Molecular Colonization of the Biosphere 18

The Architecture of Infiltration: Mapping the Human 'Plasticene'...........22

The Synthetic Gastronomy: Trophic Infiltration and the Nanoscale Siege25

The Synthetic Siege: Pathogenic Integration and the Quest for Biological Solvency 27

The Enzymatic Insurgency: Molecular Tactics for the Deconstruction of Synthetic Matter 30

The Living Scourge: Bio-Fortification and the Microbial Reclamation of the Plastisphere...33

The Mycelial Siege: Fungal Biomechanical Warfare and the Erosion of the Synthetic Frontier 36

The Mycelial Infiltration: Advanced Fungal Processing and the Mechanics of Synthetic Dissolution 38

The Thermodynamic Descent: Bio-Kinetic Pathways and the Tripartite Architecture of Synthetic Erasure 41

The Molecular Guillotine: Engineering the Enzymatic Siege of Synthetic Carbon 44

The Synthetic Reef: Bio-Architectural Colonization and the Inversion of Buoyancy 47

The Kinetic Ceiling: Stoichiometric Barriers and the Molecular Metrics of Synthetic Decay 49

The Synthetic Resurrection: Engineering the Molecular Rebirth of Industrial Waste 52

Steric Blockades and the Polyethylene Density Paradox: Engineering Ultrafiltration Synergies 56

The Adaptive Sieve: Bio-Sedimentary Filtration and the Enzymatic Erasure of Polymers 58

The Molecular Gatekeeper: Nanocomposite Membranes and the Geopolitical Siege of Synthetic Flux 62

The Membrane Paradox: Chemical Rehabilitation and the Circularity of Molecular Sieves 65

The Kinetic Barrier: Infrastructural Obsolescence and the Synergy Pyramid of Global Recovery 67

The Global Sovereignty of Purity: An Executive Strategy for the Total Neutralization of Synthetic Flux 70

The Synthetic Metabolism: Reprogramming the Microbial Secretome for Global Polymer Erasure 72

Architecting the Synthetic Scavenger: Genomic Editing and the Enzymatic Siege 74

The Kinetic Fortress: Thermodynamic Engineering and the AI-Driven Enzymatic Leap 76

The Autonomous Evolution: AI-Governed Bioreactors and the Multi-Omic Siege 79

The Bio-Architectural Reconquest: Synthetic Biology as the Terminal Solution to Polymer Recalcitrance 81

Forensic Oceanography: Establishing the Global Protocol for Synthetic Detection 83

The Molecular Pivot: Biopolymer Synthesis and the Engineering of Ecological Reciprocity 86

The Proteomic Blueprint: Carbon Flux and the Engineering of Proteinaceous Matrices 88

The Carbohydrate Scaffold: Engineering Polysaccharide Architectures for Post-Petrochemical Industrialism 90

The Autotrophic Sieve and the Forensic Logic of Decay: PHAs and the Compostability Paradox 92

The Microbial Foundry: Mycelial Composites and the Synthesis of High-Purity Bacterial Cellulose 95

The Arterial Waste-Stream: Phytological Sourcing and the Engineering of Lignocellulosic Composites 97

The Commercial Vanguard: Market Penetration and the Geopolitical Scaling of Bio-Based Polymers 100

The Jurisprudential Siege: Legislative Architectures for Global Marine Restoration 101

The Regional Bastion: Chemical Jurisprudence and the Global Mesh of Polymer Regulation 103

The Quantum Abyss: Colloidal Sovereignty and the Nanoscale Metamorphosis 105

The Behavioral and Industrial Nexus: Consumer Sovereignty and the Permitting Fortress 108

The Economic and Catalytic Lever: Extended Producer Responsibility and the CO2 Feedstock Revolution 110

The Final Synthesis: Strategic Outlook and Global Governance of the Plastisphere 113

The Behavioral Engine: Community Agency and the Deconstruction of the Plastisphere 115

The Demand-Supply Sabotage: Collective Agency as a Disruptive Force in Polymer Economics 118

The Cognitive Restoration: Pedagogical, Digital, and Grassroots Architectures for Synthetic Mitigation 121

The Upstream Mandate: Industrial Metamorphosis and the Hazard Classification of Synthetic Polymers 123

The Structural Interception: Thermodynamics, Chemical Recycling, and the Mirage of Circularity 125

The Synthetic Recession: A Multi-Generational Roadmap for Biospheric Restoration 127

The Forensic Sieve: Retrospective Remediation and the Engineering of Autonomous Ocean Recovery 130

Index 132

The Synthetic Encrustation: Navigating the Era of Universal Polymer Ubiquity

The human footprint on the biosphere has transcended mere influence, evolving into a total systemic overhaul. We have transitioned from a planet defined by biological rhythms to one encased in a managed, industrial shell. This new reality is defined by the total infiltration of synthetic polymers into every foundational layer of our world—the soil, the seas, the air, and the very cells of living organisms. These materials are no longer just external tools; they are embedded in the global fabric, appearing in every conceivable hue and chemical configuration, touching every aspect of modern existence. This omnipresence has birthed a silent narrative of health and ecological decay that we are only beginning to decipher.

To address this crisis, we must first name it, though the scientific community has struggled to find a single, unified language for these tiny invaders. At its core, we are dealing with synthetic solid particles or polymeric matrices that resist dissolution in water. These are generally categorized by a 5-millimeter upper threshold, but the danger scales down into the realm of the invisible.

The nomenclature of waste is built on a hierarchy of scale. We recognize "megaplastics" as the massive debris—nets, crates, and containers—that exceed a meter in size. Below them sit "macroplastics" and "mesoplastics," the visible fragments of a consumer society. However, the true complexity begins at the 1-micrometer mark. Here, the "microplastic" category is often split: "large microplastics" occupy the 1 to 5-millimeter range, while "small microplastics" descend toward the sub-millimeter level. At the absolute edge of detection lie "nanoplastics," particles smaller than a single micrometer. These are the most enigmatic and perhaps most dangerous of all, capable of crossing biological barriers that larger debris cannot.

Beyond size, we must classify these materials by their origin. "Primary" particles are born small—engineered for specific industrial roles, such as the resin pellets (or "nurdles") that serve as the raw feedstock for factories, or the abrasive beads found in specialized cleaners. "Secondary" particles, however, are the products of environmental violence. They are the jagged remnants of larger items, shattered by the sun's radiation, the grinding of waves, and the slow march of chemical decay.

The Geometry and Chemistry of Contamination

The physical form of these pollutants is as varied as their origins, and each shape dictates a different ecological fate. Jagged fragments, the result of structural failure in rigid plastics, pose a physical threat of internal laceration to any organism that mistakes them for food. Synthetic fibers, shed by the millions from our textiles and abandoned fishing gear, create lethal tangles within digestive tracts. Spherical pellets represent industrial leakage, while expanded foams, highly buoyant and resistant to sinking, act as long-distance vessels for contamination across oceanic currents. Finally, thin films—the ghosts of single-use packaging— suffocate surfaces and mimic prey like jellyfish.

But these particles are more than just physical hazards; they are molecular sponges. Because they are composed of complex polymers like polyethylene, polypropylene, and polyvinyl chloride, they possess a unique affinity for other toxins. They actively adsorb heavy metals, antibiotics, and persistent organic pollutants from the surrounding water, concentrating them at levels far higher than the ambient environment. Simultaneously, they leach their own internal additives—stabilizers, colorants, and endocrine-disrupting chemicals like BPA and phthalates— directly into the water and the tissues of those who ingest them.

The Petrochemical Foundation and the Cost of Permanence

The modern world is physically constructed from ancient carbon. The vast family of polymers we call plastics relies heavily on the extraction of fossil resources, claiming a significant and growing portion of global petroleum reserves. As our appetite for these materials expands, the environmental toll of their production escalates in tandem. This is not merely a waste issue; it is an extraction crisis. The processes required to harvest and refine these raw materials cause deep ecological scarring and release a cocktail of toxins into the atmosphere long before a product ever reaches a consumer. We are essentially trading non-renewable geological assets for a permanent, indestructible form of pollution.

The most harrowing manifestation of this trade-off is the invisible veil of particles now suffocating our water systems. From the deepest oceanic trenches to the most isolated mountain lakes, the hydrosphere is being re-engineered by microscopic debris. This isn't just an environmental hurdle; it is a foundational threat to the stability of global biodiversity. Our planet, already reeling from shifting climates, is now entangled in a physical web of synthetic dust. This "third dimension" of ecological collapse does more than poison local habitats; it likely alters the very biogeochemical cycles that regulate our climate, acting as a catalyst for further planetary instability.

Mapping the Global Pipeline: From Land to Sea

The narrative of marine pollution is often miscast as an offshore problem. In reality, the vast majority of synthetic debris in our oceans is terrestrial in origin. Only a tiny fraction enters the water from sea-based activities like aquaculture or shipping. The true engine of this crisis is the land-to-sea pipeline. Rivers, wind, and urban runoff act as the primary delivery systems, funneled by the sheer density of human populations living within coastal zones.

The geography of this crisis is inextricably linked to socioeconomic development. Nations experiencing rapid industrial growth without the corresponding investment in waste infrastructure become the primary contributors to the global burden. When high population density meets a lack of sophisticated recycling and landfilling systems, the result is a massive leakage of plastic into the environment. This is not a failure of a single region, but a systemic flaw in the global economic model that prioritizes production over the lifecycle of the material.

The Human Toll and Future Projections

The impact on human health is no longer a matter of speculation; it is an emerging reality. Microplastics have transitioned from the environment into our very biology, entering our systems through the air we breathe and the food we consume. The detection of these particles in human blood—found in the vast majority of tested individuals—marks a terrifying milestone. These invasive materials trigger oxidative stress and cellular damage, and their long-term persistence means they may remain within us for decades.

As we look toward the future, the trajectory is clear and concerning. Plastics are designed for endurance, not for cycles. They resist natural degradation with a ferocity that ensures they will haunt the biosphere for centuries. Without a radical shift in how we manufacture, use, and recover these polymers, we are not merely polluting the planet; we are rewriting its geological and biological code. The "engineered biosphere" is becoming a permanent monument to a material that the natural world simply cannot digest.

Mechanical Segregation: The Cascade of Exclusion

The foundational layer of isolation relies on the physical "cascade of exclusion." By passing bulk samples through a vertical hierarchy of metallic screens and glass-fiber membranes, researchers sort the chaos of the environment into mathematically distinct size brackets. Practical success in this phase demands a rejection of the very materials being studied; using polymer-based tools for filtration is a cardinal error that introduces unacceptable cross-contamination. Instead, the laboratory must rely on stainless steel and borosilicate glass—inert materials that provide a rigid, non-synthetic baseline.

However, mechanical sorting is fraught with hidden variables. The aggressive agitation required to move silt through a fine mesh can be a destructive force. Environmental plastics, embrittled by decades of solar radiation, are structurally fragile. Rough handling during sieving can shatter a single fiber into a hundred fragments, artificially inflating the final count and leading to a false perception of environmental density. To mitigate this, the modern laboratory must prioritize "gentle refinement"—utilizing vacuum-assisted filtration to move fluid through the system while minimizing the physical friction that leads to sample shattering. Furthermore, the persistent ghost of static electricity often causes the smallest particles to cling to the walls of glass funnels; a meticulous chemical wash-down is required to ensure that the "invisible fraction" is not lost to the surface tension of the equipment itself.

Electronic and Molecular Sorting: The High-Tech Frontier

As the target particle size descends into the sub-micron range, the laws of classical physics begin to fail. Small particles often lack the mass to overcome the surface tension of a salt solution, remaining trapped in a "physical limbo" neither floating nor sinking. To solve this, emerging methodologies are moving toward electromagnetic and chemical affinity. Electrostatic separators, borrowing technology from the heavy mining industry, exploit the insulating properties of polymers. By subjecting a dry sample to a high-voltage field—often exceeding 30 kilovolts—researchers can physically "flick" the synthetic particles away from the conductive mineral grains, bypassing the need for density-based liquids entirely.

Parallel to this, the industry is experimenting with froth flotation, where air bubbles are utilized as "biological taxis." By manipulating the hydrophobicity of the plastic surface, researchers can cause bubbles to latch onto the polymer and carry it to the surface. However, this method remains notoriously unstable; the physical inconsistencies of the bubbles often lead to high loss rates, making it more suitable for industrial recycling than for precise environmental forensics.

Deep Analysis and Practical Strategy

The most critical oversight in contemporary isolation is the failure to account for "bio-encrustation." In the wild, plastics do not exist as clean, smooth surfaces; they are colonized by microbial biofilms that act as a biological anchor, dragging lightweight plastics down into the "sink" of the sediment. If these biofilms are not chemically dissolved prior to density separation, the researcher is essentially measuring the weight of the life on the plastic rather than the plastic itself. The practical advice for the next generation of researchers is to prioritize aggressive enzymatic or oxidative digestion—using hydrogen peroxide or specialized proteins—to "clean" the particle before attempting to sort it. This ensures that the buoyancy of the particle reflects its true chemical identity, not its biological passenger.

Future Predictions: The Rise of the "Lab-on-a-Chip"

I predict that the next ten years will see the total obsolescence of the static sieve and the salt brine. The future of microplastic isolation lies in Acoustic Microfluidics and Pinched Flow Fractionation. We are moving toward a reality where a raw water sample is pumped through a microscopic, etched channel where ultrasonic waves act as invisible filters. These sound waves will "nudge" particles of different masses and densities into separate outlets with a precision that no mechanical screen can match.

Furthermore, the integration of Machine-Learning Enhanced Cytometry will allow for the simultaneous isolation and identification of particles. Rather than spending weeks manually counting fibers under a microscope, researchers will utilize high-throughput fluid systems that take a high-speed photograph of every particle, identify its polymer type via an onboard neural network, and sort it into a reservoir for further study—all in a fraction of a second. The bottleneck of the "analytical purgatory" will finally be broken, allowing for a real-time, global mapping of the synthetic crisis as it unfolds.

Gravimetric Defiance: Navigating the Buoyancy Threshold

When dealing with dense mineral matrices, such as coastal clays or deep-sea silts, mechanical sieving is often insufficient. Here, the protocol shifts to the manipulation of fluid dynamics through density-based flotation. The premise is simple: create a brine so heavy that the synthetic world floats while the geological world sinks. Because the specific gravity of most commercial polymers is significantly lower than that of quartz or carbonate, researchers can engineer a "buoyancy threshold" to skim the synthetic supernatant from the mineral dross.

The choice of the brine, however, is a study in compromise. While a simple sodium chloride solution is cost-effective and safe, its low density fails to capture the "heavy" polymers—PVC and PET—that represent a massive portion of the industrial footprint. To capture the full spectrum, the laboratory must turn to heavy salts like zinc chloride or sodium iodide. Here lies the practical dilemma: these substances are often toxic, corrosive to equipment, and prohibitively expensive. The reliance on these heavy brines creates a barrier for researchers in developing regions, leading to a global data gap where the most polluted areas are often the least capable of performing a comprehensive census of their own environment.

The Final Convergence: A Unified Theory of Detection

As we synthesize the methodological imperatives of the last decade, it is clear that the future of monitoring lies in Global Calibration. The current landscape of ad-hoc university studies is insufficient for the demands of international law. We require a "Universal Metric" where a sample extracted from the Arctic can be directly compared to one from the South China Sea.

The Sampling Nexus Field extraction remains the most vulnerable phase of the cycle. Whether utilizing surface-skimming gliders like the Manta trawl or deep-sea mechanical grabbers, the choice of mesh size is the defining variable. Researchers must balance the need to capture the exponentially more dangerous "nano-fraction" against the physical reality of net clogging.

Practical Advice: Always utilize a mechanical flowmeter to record the exact volume of water processed; without this metric, the final data is merely qualitative and cannot be used for the hydrodynamic modeling required for global policy.

Chemical Refinement and the Saponification Standard The removal of organic "noise" is the final hurdle before the laser meets the polymer. While acid digestion is powerful, it is too destructive for the fragile polymers of the 21st century. The modern standard has shifted toward potassium-based saponification (KOH) for biological tissues and controlled oxidation (H2O2) for environmental sediments.

Prediction: I predict that the next five years will see the mandatory adoption of ISO-Certified Clean-Room Protocols for all microplastic research. Data that does not include procedural blanks, environmental controls, and a verified "Natural Fiber Log" will be summarily rejected by the scientific community. We are moving toward a reality where the quality of the laboratory's air is as important as the quality of the sample itself.

The Forensic Identification Crisis: Decoding the Synthetic Signature

The identification of environmental microplastics represents the ultimate forensic challenge in contemporary oceanography. Because these particles are not a single material but a chaotic mosaic of varying polymers, degradation states, and industrial additives, simply counting them is an exercise in futility. To provide data that can drive global policy, researchers must move beyond mere tallying and achieve a definitive chemical "verdict" for every fragment. This analytical journey begins with optical reconnaissance and culminates in the total molecular incineration of the sample to reveal its hidden chemical history.

Analytical Conclusion: The Future of the Global Census

 The "Synthetic Encrustation" of our planet is a problem of scale and invisibility. Our tools have finally reached the level where we can see the invisible, but our methods must now reach the level where we can trust what we see. By integrating rigid QA/QC frameworks with advanced molecular identification, we are building the evidence base for a planetary intervention.

The ultimate goal of this methodological evolution is the transition from "Environmental Monitoring" to "Industrial Accountability." When we can definitively trace a 10-micrometer fragment of weathered PVC back to its industrial point of origin—accounting for its chemical additives and its history of oxidation—the era of anonymous pollution will end. The "Sterile Bastion" of the laboratory is the first step in reclaiming a biosphere that is no longer defined by the indestructible remains of our own production.

Analysis of Human Error: Even with staining, human observation is plagued by subjectivity and fatigue. Statistical audits of trained researchers reveal a variance in detection rates of up to 40% for the same sample. We are reaching a "Human Ceiling" in microplastic quantification.

Prediction: Within the next three to five years, manual optical sorting will be entirely replaced by AI-Driven Machine Vision. Neural networks, trained on petabytes of imagery showing degraded polymers in every state of decay, will provide a standardized, bias-free count at a speed and accuracy level that no human team can match.

Molecular Incineration: Pyr-GC-MS and the Additive Profile

For the most complex samples, where light-based methods are blinded by surface contamination, the laboratory turns to Thermal Dissection. Pyrolysis-Gas Chromatography-Mass Spectrometry (Pyr-GC-MS) does not look at the plastic; it destroys it. By heating the sample to 800°C in an inert atmosphere, the polymer backbone is shattered into a gaseous cocktail.

This "Brute Force" method is the only way to reveal the hidden architecture of the plastic—the toxic plasticizers, flame retardants, and antioxidants that were baked into the polymer during its industrial birth. While this data is invaluable for toxicological risk assessments, the cost is the total destruction of the evidence. You cannot count particles with Pyr-GC-MS; you can only measure the mass of the synthetic ash.

The Nanoscale Frontier: Hyphenated Evolution

As the crisis scales down into the "Nano-Realm," we are entering an era of Hyphenated Technologies. Standard microscopes cannot see a nanoplastic, and standard lasers cannot resolve its chemistry. The future lies in tools like AFM-IR (Atomic Force Microscopy-Infrared), which uses a physical needle to "feel" the topography of a nanoplastic while simultaneously firing a laser to identify its molecular structure.

The ultimate evolution is the TGA-FTIR-GC-MS—a massive, interconnected analytical chain that monitors a sample as it burns, identifying every chemical peak in real-time with zero margin for error.

Concluding Prediction: We are moving toward a "Real-Time Synthetic Census." The current model of manual field collection followed by months of lab work is dying. I predict that the end of this decade will see the deployment of In-Situ Autonomous Spectrometers—robotic buoys that filter, identify, and report the chemical typology of microplastics in real-time via satellite. The "Analytical Bottleneck" will be solved not by better labs, but by eliminating the lab entirely and moving the forensic inquest directly into the water column.

The Synthetic Trophic Shift: Dynamics of Polymer Integration and Marine Stratification

The biological landscape of our planet is currently undergoing a permanent structural re-alignment. The widespread infiltration of microscopic synthetic matter is no longer merely an external environmental "issue"; it is a foundational shift in how matter and energy move through the biosphere. When a foreign, indestructible element is introduced into an established ecosystem, it does not remain a passive observer. Instead, it forces a destructive synergy with both the physical environment and the living communities it invades. Microplastics represent the final stage of this invasion—a material that has successfully bypassed the natural evolutionary defenses of living organisms to become a permanent, internal component of the global food web.

The Fragmentation Cascade: From Industrial Debris to Molecular Infiltration

The journey of synthetic matter begins in the terrestrial hubs of human civilization. It is a stark reality of modern physics that the vast majority of our production is gravity-fed. Because the world's oceans sit at the lowest topographical points on the planet, they act as the ultimate, inescapable reservoir for every discarded polymer. This land-to-sea pipeline is powered by a relentless hydrological connection: mountain streams, industrial drainage, and urban stormwaters all serve as the high-speed conveyor belts that deliver 75% of the total marine plastic load. Once these materials reach the oceanic basins, they enter a state of "suspended permanence," where they will remain for centuries, shielded from natural decay by their highly engineered chemical resilience.

To manage this crisis, we must recognize the "Fragmentation Cascade." Debris is classified not by its utility, but by its scale of destruction. At the top sit Megaplastics and Macroplastics—the visible ghosts of the industrial world, such as abandoned nets and shipping containers. Below these are Mesoplastics, the intermediate fragments that serve as the primary source for the true ecological invader: Microplastics.

These particles, defined by a sub-five-millimeter threshold, are rarely born small. Instead, they are the product of environmental violence— shattered by the sun's ultraviolet radiation, ground down by wave friction, and weighed down by biological growth. This process of infinite fragmentation ensures that even if we ceased all production today, the number of individual particles in the sea would continue to grow exponentially for decades.

Hydrodynamic Stratification and the Buoyancy Paradox

Once a polymer enters the marine biome, its ecological fate is dictated by a single physical variable: density. This creates a highly stratified "Synthetic Column" that reaches from the sunlit surface to the crushing depths of the abyss.

1. **The Surface Residents** Lightweight polymers, such as the polyethylene and polypropylene used in universal packaging, possess a specific gravity lower than seawater. These materials are destined to ride the great oceanic gyres, forming vast, floating synthetic "meadows" that block light and interfere with surface-dwelling life.

2. **The Benthic Graveyard** Conversely, high-density industrial polymers like PVC, PET, and heavy polyamides (Nylon) are doomed to sink. They descend through the water column, eventually blanketing the benthic floor and the lightless zones of the deep sea. This creates a "smothering effect" on the sediment-dwelling organisms that form the base of the deep-sea food web.

Analysis of the Buoyancy Shift: It is a critical error to view these positions as static. The biological process of "encrustation" or biofouling—where algae, bacteria, and barnacles colonize the plastic surface—fundamentally rewrites the particle's physics. Life itself acts as a biological anchor; a piece of plastic that floats today will inevitably sink tomorrow once its biological load overcomes its natural buoyancy. In this way, no layer of the ocean is safe; microplastics are dynamically distributed through every stratified sub-zone of the planet.

Practical Advice for Structural Mitigation

For those designing the next generation of materials and policy, the focus must shift from "recyclability" to "programmed obsolescence."

- **Enzymatic Triggers:** We should prioritize the development of polymers embedded with dormant, heat-sensitive, or pressure-sensitive enzymes. These would remain inactive during the product's shelf-life but trigger a rapid molecular breakdown once exposed to the specific pH or pressure of the deep-sea environment.

- **Surface Modification:** Developing "anti-fouling" coatings for unavoidable industrial plastics could prevent the vertical migration of debris, keeping it in the surface layers where mechanical recovery is at least theoretically possible.

Predictions: The Hybrid Biosphere

I predict that the next fifty years will witness the rise of a "Hybrid Biosphere." We are already seeing the first signs of microbial life evolving to utilize plastic as a primary substrate and carbon source. We are moving toward a reality where the "Plastisphere" is not an external pollutant, but a new, synthetic niche that will host unique evolutionary lineages.

The Synthetic Canopy: Thermodynamic and Biogeochemical Disruption of the Oceanic Engine

The global ocean is far more than a biological repository; it is the primary thermal and chemical regulator of the biosphere. While terrestrial systems are often the focus of climate discussions, the true stability of the planet is maintained by the deep thermodynamics of the marine environment—a system powered by surface-level solar energy and sustained by abyssal chemosynthetic networks. The infiltration of microplastics is not merely a localized toxicity issue; it is a systemic "clogging" of the planet's largest engine. By introducing a persistent, foreign particulate phase into the water column, we are effectively altering the "atmospheric" chemistry of the sea, creating a subsurface smog that destabilizes the very mechanisms that prevent runaway climate volatility.

Photonic Starvation: The Optical Impedance of Synthetic Films

The most immediate threat to the oceanic engine is the disruption of the "Photonic Pipeline." In a pristine state, approximately 80% of solar radiation is absorbed within the upper 10m of the water column, fueling the primary production that drives all marine life. However, the accumulation of high-density synthetic debris on the surface creates an

"Artificial Canopy" that fundamentally re-engineers the water's optical properties.

Metabolic Suffocation and the Synthetic Bio-Colony

The oxygen dynamics of the marine environment are precarious. With an average concentration of only 9 ml/L—roughly 4% of the terrestrial atmospheric level—marine organisms operate on a tight metabolic budget. Microplastic pollution introduces a catastrophic "Metabolic Tax" on this system.

As particles circulate, they provide a stable, high-surface-area substrate for microbial colonization. This "Synthetic Bio-Colony" transforms the particles into mobile sites of intense metabolic activity. Billions of microorganisms proliferating on these plastic surfaces consume dissolved O_2 at an accelerated rate through decomposition and respiration. This results in the creation of localized "Hypoxic Micro-Zones." For higher organisms, this reduced oxygen availability leads to depressed energetics and lethal physiological stress. We are effectively witnessing the birth of a "Suffocating Sea," where the energy required for survival exceeds the oxygen available in the environment.

The Thermal Blanket: Carbon Sink Sabotage and Heat Trapping

The ocean acts as the planet's most vital carbon sink, sequestering CO_2 through both physical and biological mechanisms. The mass death of phytoplankton—driven by the shading effects described above—drastically reduces the efficiency of this sequestration. This "Carbon Pump Sabotage" creates a terrifying climate feedback loop: as the ocean loses its ability to absorb carbon, atmospheric warming accelerates, which in turn increases the degradation rate of plastics, releasing even more microplastics into the water.

Furthermore, a localized thermal mechanism exists that acts like a "Synthetic Blanket." Dense layers of floating microplastics obstruct the natural convective loss of heat from the ocean surface. By absorbing solar radiation and trapping the heat generated by internal metabolic functions, these plastic layers act as an insulating skin.

Deep Analysis: This insulation causes localized "Lethal Heat Spikes" in the upper mixing layer. Unlike the broad patterns of global warming, these synthetic-induced heat traps are sudden and concentrated, capable of cooking delicate larval stages of fish and coral before they can reach maturity.

Practical Advice for Environmental Forensics

To accurately monitor these emerging "Lethal Heat Traps" and "Hypoxic Micro-Zones," we must move beyond simple count-based sampling.

- **Thermal Mapping:** Researchers should utilize high-resolution infrared drone surveys to identify "Thermal Anomalies" on the ocean surface that correlate with high plastic density.

- **Dissolved Oxygen Profiling:** Sampling protocols should prioritize real-time O2 sensors attached to surface-skimming nets to measure the "Metabolic Depletion" caused by the synthetic bio-colony in situ.

Predictions: The Great Oceanic Albedo Shift

I predict that within the next three decades, we will witness the Great Oceanic Albedo Shift. As the surface concentration of plastics continues to grow, the reflective properties of the ocean will change enough to be visible from deep space. While this might slightly reduce local water temperatures by reflecting sunlight, the trade-off will be the total collapse of the biological carbon pump.

Furthermore, I predict the emergence of Plastic-Induced Dead Zones (PIDZs). These will be regions of the ocean where oxygen levels drop to zero not because of nutrient runoff, but because the "Metabolic Tax" of the synthetic bio-colony has stripped the water of its life-sustaining gases. The future of the ocean is one of "Inorganic Smog," where the thermodynamic stability of the planet is traded for a permanent, synthetic film that prevents the sea from breathing.

The Biological Trojan: Synthetic Bio-Accumulation and the Erasure of Species Boundaries

The contemporary biosphere is no longer a closed system of organic nutrients. We have entered an era of "Synthetic Nutrient Hijacking," where the very pathways designed to sustain life—the food chains and atmospheric cycles—have been repurposed as delivery systems for industrial waste. The infiltration of microplastics into the human and animal soma is not a passive event; it is an active molecular occupation. By bridging the gap between the inert and the living, these particles act as "Biological Trojans," smuggling a lethal payload of petrochemicals across the most sacred cellular barriers.

The Trojan Mechanism: Synergistic Toxicity and Endocrine Sabotage

The danger of a microplastic particle is not merely its physical presence, but its function as a high-capacity chemical vessel. Once a particle is inhaled as atmospheric dust or ingested through a contaminated water source, it begins a two-phase chemical assault on the host.

1. **Internal Leaching:** Modern polymers are rarely pure; they are complex matrices impregnated with Bisphenol-A (BPA), Phthalates, and halogenated flame retardants designed for durability. Once trapped in the acidic, warm environment of the human digestive or respiratory tract, these additives unbind from the polymer and flood the bloodstream.

2. **Sorption Magnet:** During its time in the ocean, the hydrophobic surface of the plastic aggressively adsorbs ambient toxins—specifically DDT, PCBs, and heavy metals like Hg (Mercury) and Pb (Lead).

The Upward Spiral: Trophic Transfer and Nutrient Corruption

The "First Law of Ecology" dictates that no element exists in isolation. In a contaminated world, the food web becomes a "Biological Multiplier." This process of Trophic Transfer ensures that the smallest synthetic error at the base of the ecosystem is magnified into a lethal crisis for those at the top.

The cycle begins at the second trophic level, where filter-feeding zooplankton mistake microscopic shards for natural phytoplankton. As these primary consumers are hunted, the synthetic load is passed upward, concentrating with every kill. This is not a theory but a documented reality reaching the most extreme points of our planet. Microplastics have been recovered from the digestive systems of predatory organisms in the mid-water column and, most hauntingly, from amphipods inhabiting the lightless floor of the Mariana Trench. Even at the Earth's deepest point, the synthetic fibers of human textiles have become part of the biological record.

The Terrestrial-Aquatic Loop This contamination is not seaward-bound. Predators such as coastal bears, sea birds, and amphibians act as mobile bridges, carrying the marine plastic load back into the heart of terrestrial ecosystems. When a land-based predator consumes an anadromous fish, the "Synthetic Food Chain" is closed, ensuring that no habitat, regardless of its distance from the coast, remains pristine.

Practical Advice for Clinical and Environmental Intervention

To combat this systemic infiltration, we must move beyond general environmentalism toward Molecular Forensics and Material Reform.

- **Clinical Screening:** Medical professionals should move toward "Body Burden Analysis" for patients in high-risk zones, screening not just for heavy metals but for the presence of specific plasticizer metabolites in blood and urine.

- **Hydrophobic Material Design:** We must prioritize the development of "Low-Sorption Polymers" that lack the chemical affinity for Persistent Organic Pollutants (POPs), reducing their ability to act as chemical sponges during their environmental life cycle.

- **Consumer Awareness:** Given that seafood is the primary return-vector for marine plastics to human biology, regulatory bodies should mandate "Synthetic Load Labels" on commercial fish products, similar to mercury warnings.

Predictions: The Era of Genomic Weathering

I predict that within the next forty years, we will recognize a new clinical phenomenon: Genomic Weathering. This will be defined as the transgenerational damage caused by the constant presence of EDCs in the human germline. We are likely to see a permanent shift in human baseline health, where "natural" hormonal balance becomes a historical anomaly.

Furthermore, I predict the rise of Atmospheric Polymerization. The loop will not only close through the food we eat but through the air we breathe. As violent coastal storm systems and aerosolized ocean sprays continue to launch marine microplastics into the high atmosphere, we will see "Synthetic Rain" falling over deep terrestrial agricultural lands. This will create a state of "Universal Saturation," where the distinction between a "marine" and "terrestrial" pollutant is entirely erased. The future of the biosphere is one of a single, unified, synthetic metabolism—a world where the plastic we discarded yesterday becomes the cellular building block of the life born tomorrow.

The Polymer Terminal: Geopolitical Inertia and the Molecular Colonization of the Biosphere

The geopolitical trajectory of the 21st century suggests that our reliance on synthetic polymers has reached a state of industrial inertia. Far from a decline, the global appetite for both single-use and high-performance plastics is projected to surge as these materials become foundational to emerging sectors like renewable energy infrastructure, advanced medical robotics, and hyper-automated logistics. We are not merely living in a plastic age; we are witnessing the total integration of a synthetic metabolism into the global economy. As coastal populations swell and hyper-urbanization accelerates, the baseline for plastic emissions will move from a crisis to a permanent environmental constant. Unless there is a violent shift toward programmed molecular disassembly—where materials are engineered to self-destruct at the atomic level—the biosphere faces a state of irreversible biome collapse.

The Myth of Biotic Salvation: Thermodynamic Realities

A popular, albeit speculative, narrative in contemporary science focuses on "Biotic Salvation"—the idea that naturally evolving or genetically tailored "super-strains" of bacteria and fungi will eventually digest the petrochemical problem. However, a deep thermodynamic analysis suggests this optimism is misplaced. While laboratory settings show promise, the physical reality of the marine environment is one of cryogenic preservation.

In the high-latitude temperate and polar regions, where the water is dark and near-freezing, enzymatic cleavage and microbial metabolism slow to a crawl. The sheer kinetic energy required to break a carbon-carbon polymer backbone is rarely available in these deep-sea sinks. Consequently, we must accept a permanent accumulation scenario: the rate of synthetic influx from terrestrial sources will continue to outpace the planet's biological "digestive" capacity by several orders of magnitude, rendering current bioremediation efforts statistically insignificant on a planetary scale.

The Nanoscale Soup: From Fragmentation to Infiltration

The environmental fate of plastic does not end with the micro-particle. We are moving toward a terminal fragmentation state. Driven by surface photo-oxidation and the aggressive, yet incomplete, grazing of microbes, microplastics are being shattered into a sub-micron "soup." This transition from micro to nano represents a shift from a pollution problem to a genomic threat.

Because they exist at the nanoscale, these particles achieve a state of intracellular infiltration. They no longer simply obstruct the gut; they bypass the blood-brain barrier and penetrate the cellular membrane, physically disrupting organelle function. We are moving toward a future where the baseline survival behavior of entire species is altered by the presence of synthetic interference within their central nervous systems. Paradoxically, the only silver lining is that the extreme surface-to-volume ratio of nanoplastics may make them slightly more vulnerable to terminal oxidation, potentially leading to a faster ultimate molecular breakdown— provided the influx of new material is halted.

The Unified Theory of Contamination: A Systemic Summary

The "Ecological Paradigm" of the 21st century is defined by the absolute saturation of the hydrosphere. The lifecycle of a plastic product is a journey from industrial utility to biological weapon. These particles act as sorption magnets, concentrating lethal persistent organic pollutants (POPs) and neurotoxic heavy metals like Pb (Lead) and Hg (Mercury) on their surfaces.

The Multi-Dimensional Impact:

- **Thermodynamic Sabotage:** By altering light penetration and surface albedo, plastics disrupt the biological carbon pump, effectively "suffocating" the ocean's ability to regulate the global climate.

- **Internal Poisoning:** Within the human and animal body, the breakdown of the polymer releases a "Trojan Payload" of endocrine-disrupting plasticizers, BPA, and carcinogenic flame retardants. The result is a global escalation in metabolic reconfiguration, reproductive sterilization, and degenerative neurological decay.

- **Biological Erasure:** The distinction between the "Marine" and "Terrestrial" world has been erased. Through the migratory patterns of coastal birds and the consumption of commercial seafood, the marine plastic load is being pumped back into the human heartland.

Practical Advice for Industrial and Policy Reform

To sever the link between modern life and synthetic decay, the global community must move beyond the "Recycling Myth" toward structural material reform:

- **The Zero-Sorption Mandate:** Policy must prioritize the development of polymers that lack the chemical affinity for ambient environmental toxins, preventing them from acting as "Trojan Horses" during their environmental life cycle.

- **Histological Audits:** Regulatory bodies should move toward mandatory histological auditing of commercial fish stocks, screening not for the presence of plastic, but for the presence of cellular-level "nanoplastic signatures."

- **Standardized Molecular Forensics:** We must bridge the "epistemological void" by funding advanced spectroscopic and hyphenated technologies (TGA-FTIR-GC-MS) capable of tracking the sub-micron behavior of these particles in real-time.

Predictions: The Era of Molecular Trespass

I predict that the next thirty years will be defined by the emergence of molecular trespass lawsuits. As the detection of nanoplastics in human blood and fetal tissue becomes a routine clinical metric, we will see a global legal shift where industrial producers are held liable for the "occupational colonization" of the human body.

Furthermore, I predict a global albedo shift. The volume of plastic on the ocean surface will reach a point where it measurably increases the planet's reflectivity. While this may temporarily mask the effects of global warming, the trade-off will be the total collapse of the marine photosynthetic engine. The future of the ocean is one of a "synthetic dead zone"—a world where the water is saturated with an inorganic smog that prevents the sea from breathing.

The ultimate terminal fate of our civilization will be determined not by our technology, but by our ability to manage the indestructible remains of our own production. We are rewriting the geological and biological code of the planet; the question is whether any organic life will be left to read it.

The Architecture of Infiltration: Mapping the Human 'Plasticene'

The global geography of waste is no longer defined by localized landfills, but by massive, autonomous structures known as synthetic archipelagos. The most notorious of these—the North Pacific's vortex of debris—has evolved into a sprawling, non-natural mass. Within this swirling gyre, billions of fragments have coalesced into a density so profound that it is frequently characterized as the planet's newest geological feature. While the bulk of the mass consists of larger debris, the numerical majority— roughly 94% of all individual items—is comprised of microscopic shards. This universal saturation of the ocean has effectively ushered the biosphere into the Plasticene, a novel historical epoch where the distinction between the organic environment and synthetic manufacturing has effectively vanished.

The Geography of Internalized Waste

In the Plasticene, human physiology has become the terminal destination for industrial polymers. Unlike historical pollutants with volcanic or mineral analogues, these synthetic materials are 100% anthropogenic, designed for the ephemeral convenience of a consumer society but possessing a geological lifespan. Because our global infrastructure—from the logistics of the food chain to the carpeting in our homes—is built upon degrading polymers, human exposure is a mathematical certainty.

Empirical surveys have identified microplastics in every foundational element of human life: they are present in the municipal water we drink, the salt and sugar we use for seasoning, the honey harvested from remote regions, and even the "pristine" glaciers of the Arctic. We have moved beyond an "environmental crisis" into a state of biological occupation. Modern domestic epicenters, specifically the kitchen and bathroom, serve as primary sites of infiltration, where the mechanical friction of opening a package or the thermal degradation of a cooking utensil releases a constant stream of sub-micrometric shards into the immediate vicinity.

The Mechanics of Exposure: Inhalation and Ingestion

Human beings are currently subject to two relentless physiological conduits of contamination: the continuous inhalation of synthetic atmospheric dust and the direct gastrointestinal absorption of tainted nutrients and water.

1. **The Respiratory Siege** The air within modern buildings is often an order of magnitude more contaminated than the air outdoors. This is driven by the "textile shedding" of synthetic garments (polyester, nylon, acrylic) and the slow erosion of interior furnishings. These aerosolized particles are trapped by the mucus membranes and either absorbed into the lungs or swallowed, leading to a chronic, daily internal accumulation.

2. **The Dietary Vector** The most insidious dimension of the crisis is the integration of polymers into the foundational architecture of the global food web.

Deep Analysis: Developmental Sabotage The highest risk is concentrated in the most vulnerable developmental windows. The medical infrastructure itself is a source of exposure; sterile hospital environments are saturated with polymer-based devices. Most concerning is the ubiquitous practice of feeding infants using polypropylene bottles. When these bottles are subjected to the high temperatures required for sterilization, they can shed millions of microplastic fragments into the liquid, ensuring that a child's first nutritional intake is heavily laced with synthetic debris.

Practical Advice for Physiological Defense

While total avoidance is impossible in a polymer-based civilization, several strategic shifts can drastically reduce the individual body burden:

- **Material Transition:** Aggressively replace synthetic home textiles (fleece, polyester carpets) with natural alternatives like cotton, wool, or linen to minimize the "textile smog" in your living space.

- **Air Purification:** Utilize HEPA-filtered vacuum units and air scrubbers to capture the aerosolized synthetic dust that standard ventilation systems miss.

- **Thermal Awareness:** Never heat food or liquids in plastic containers, regardless of "microwave-safe" labeling. The combination of heat and friction is the primary catalyst for molecular shedding.

- **Water Filtration:** Move away from commercially bottled water—which contains higher concentrations of micro-debris due to bottle degradation—in favor of high-quality glass-housed filtration systems.

Predictions: The Surveillance of the Invisible

I predict that within the next decade, we will witness the mandatory integration of polymer sensors. Standard municipal air quality reports will move beyond PM2.5 and PM10 to include synthetic fiber density (SFD) metrics. These will become as heavily regulated as carbon emissions, particularly in high-density urban zones.

Furthermore, I predict the emergence of clinical plastic burden testing. Similar to how we currently screen for cholesterol or heavy metals, future medical diagnostics will include a standardized polymer load score. Insurance premiums and public health guidelines will eventually be dictated by an individual's internal plastic count. The current "black box" of human health in the Plasticene will be opened by new histological techniques, revealing that many chronic, idiopathic pulmonary and metabolic diseases are, in fact, the result of a lifetime of synthetic weathering within our own tissues.

The Synthetic Gastronomy: Trophic Infiltration and the Nanoscale Siege

The human diet has undergone a silent, structural re-engineering. What was once a system of organic nutrient cycling has been transformed into a terminal destination for industrial waste. Because the global hydrosphere acts as the ultimate, inescapable sink for the planet's synthetic debris, the marine-to-human dietary axis has become the primary conduit for polymer re-entry into the human soma. We are no longer merely consuming food; we are participating in a state of synthetic gastronomy, where every meal processed through a polymer-based infrastructure carries a measurable payload of industrial matter.

The Marine-Dietary Axis: From Benthic Sinks to Human Tissues

The marine environment is the most concentrated site of this contamination. While the ocean is the most visible vector, the modern supermarket has created a parallel terrestrial infrastructure of contamination. The statistical risk of exposure escalates exponentially the moment a food item enters the processing pipeline. Highly processed and aggressively packaged goods are not merely surrounded by plastic; they are chemically integrated with it.

The Container-to-Content Translocation: Thermal and Chemical Triggers

The most insidious dimension of dietary exposure is the packaging paradox, where the very material designed to protect food acts as an active contaminant. This is a process of molecular leaching, where the physical walls of a container shed micro-debris directly into the food matrix. This translocation is violently accelerated by three primary catalysts:

- **The Thermal Trigger:** The application of heat, particularly through microwaving, causes the polymer matrix to lose its structural integrity, "sweating" additives directly into the meal.

- **The Acidity Factor:** High-acid foods (like tomato-based sauces) chemically erode the surface of the plastic.

- **Lipid Affinity:** Because many toxic additives are lipophilic, they have a natural chemical affinity for fatty foods, migrating from the packaging into the nutrient at a high rate.

The list of infiltrating polymers is extensive, including PE, PET, PUR, PS, PVC, PP, PA (Nylon), and PMMA. These materials serve as the delivery system for endocrine saboteurs—specifically BPA and phthalates—which disrupt the body's internal signaling the moment they pass through the gastrointestinal lining.

The Sub-Micron Ghost: Nanoplastic Saturation

As our analytical resolution increases, we encounter the sub-micron ghost. Nanoplastics—particles ranging strictly from 1 to 100 nm—represent the ultimate terminal stage of polymer decay. Currently, we are operating in an analytical black hole; because these particles are too small for conventional light-based detection, we have almost zero empirical data on their true concentration in the global food supply.

However, the toxicological implications are terrifying. Unlike larger microplastics, which are often physically too large to exit the digestive tract, nanoplastics possess quantum-level bioavailability. They are small enough to breach the selective intestinal barrier, enter the vascular system, and potentially cross the blood-brain barrier. This allows for the direct, molecular-level poisoning of the central nervous system—a threat that traditional food safety regulations are entirely unequipped to manage.

Deep Analysis: The Eradication of the 'Organic' Baseline

We must acknowledge that the concept of "pure" or "organic" food is becoming a historical relic. Even if the soil is pristine, the atmospheric fallout of synthetic fibers and the contamination of the water table ensures a baseline level of saturation. The true danger lies in the synergistic cocktail effect: the polymer is never just plastic; it is a sponge for ambient heavy metals and persistent organic pollutants (POPs) that it collects during its journey through the environment, delivering them directly to the human cellular membrane.

Predictions: The Forensic Revolution in Food Safety

I predict that within the next decade, we will witness a global regulatory paradigm shift. As hyphenated analytical forensics—specifically field-flow fractionation coupled with advanced mass spectrometry—becomes the standard, we will finally see the sub-micron ghost. This data will likely reveal that our previous estimates of plastic ingestion were underestimated by orders of magnitude.

Furthermore, I predict the rise of polymer-free certification. Similar to the non-GMO or organic movements, we will see a high-tier food market based on verified synthetic-free supply chains. This will eventually lead to a bifurcated food system: a lower-tier industrial diet saturated with polymers, and a premium biogenic diet for those who can afford the infrastructure required to keep the synthetic world at bay. Ultimately, the medical community will begin to recognize synthetic bio-accumulation as a primary driver of idiopathic metabolic and neurological disorders, forcing a total rewrite of public health guidelines.

The Synthetic Siege: Pathogenic Integration and the Quest for Biological Solvency

The human organism is no longer a purely biological entity; it has become a porous vessel for the industrial age. The transition from the Holocene to a polymer-saturated epoch is not merely a geological observation but a cellular reality. We are currently witnessing a synthetic siege, where the boundary between the internal human environment and the external industrial waste stream has effectively dissolved. This is not a localized contamination event, but a total immersion of the species within a fragmented, indestructible matrix that infiltrates the air we breathe, the fluids we consume, and the very skin that supposedly protects us.

The Multi-Vector Occupation: From Inhalation to Trans-dermal Breach

The architecture of human exposure is a complex, multi-tiered system of infiltration. Our respiratory systems act as high-volume filters for an atmosphere laden with aerosolized microfibers—the synthetic smog shed from textiles, furnishings, and degrading infrastructure. Simultaneously, our hydration cycles—both municipal and commercial—deliver a constant baseline of polymers directly to the gastrointestinal lining.

However, the most insidious vectors are those often ignored in standard toxicological models:

- **The Developmental Inception:** The very infrastructure of early human care has been compromised. The use of high-performance polymers in infant feeding systems, when subjected to the thermal stress of sterilization, acts as a primary delivery mechanism. This ensures that the first nutritional intake of a developing organism is heavily laced with sub-micron shards, potentially altering metabolic and immunological baselines before the child reaches maturity.

- **The Dermal Veil:** We must acknowledge the trans-dermal breach. The intentional inclusion of primary microbeads (strictly < 1 mm) in personal care products—ranging from sunscreens and lipsticks to aerosolized deodorants—creates a direct pathway for systemic entry across the skin barrier. This bypasses the digestive system entirely, allowing synthetic matter to enter the lymphatic and vascular systems through passive absorption.

The Pathogenic Intersection: Antibiotic Resistance and the 'Plastisphere'

As we look toward the biological remediation of this crisis, we encounter a terrifying evolutionary side effect. Microplastics in the marine environment do not exist as inert debris; they function as synthetic reefs or the plastisphere. These surfaces serve as high-density meeting grounds for diverse bacterial colonies, facilitating a phenomenon known as horizontal gene transfer (HGT).

Deep Analysis: The Rise of the Polymer Superbug The plastisphere acts as a mobile laboratory for the evolution of antibiotic resistance. By providing a stable substrate for disparate bacterial species to interact, microplastics accelerate the sharing of resistance genes. This carries grave implications for global medicine: the same plastic particle that carries a toxic chemical payload may also be transporting a colony of multi-drug-resistant pathogens. We are not just dealing with a pollution problem; we are dealing with a pathogenic vector problem that could fundamentally devalue current antibiotic therapies.

Practical Advice: Strategies for Systemic and Individual Mitigation

To navigate this occupation, we must move beyond awareness toward active bio-defense:

- **The Thermal Decoupling:** To reduce the shedding of polypropylene and polyethylene, all polymer-based containers must be decoupled from heat sources. Thermal stress is the primary catalyst for molecular fragmentation; transitioning to inert materials (glass, medical-grade silicone, or stainless steel) for heated liquids is a non-negotiable step in reducing the body burden.

- **Filtration Reform:** Municipalities must prioritize the installation of membrane-based ultra-filtration (< 0.1 μm) to capture the sub-micron fraction that current sand-and-carbon filters miss.

- **Atmospheric Hygiene:** In indoor environments, high-exchange HEPA filtration is the only effective defense against the textile smog generated by fast-fashion fabrics and synthetic carpets.

- **Biochemical Boycott:** Consumers should aggressively identify and reject products containing polymeric abrasives. Any ingredient list featuring polyethylene, polypropylene, or nylon in a topical application represents a direct risk of trans-dermal infiltration.

Predictions: The Bio-Digital Search for Solvency

I predict that the next two decades will be defined by the race for genetically engineered remediation. We are moving toward a reality where super-strains of hyper-metabolizing microbes will be deployed to mine the oceans of their synthetic load. However, the thermodynamic reality of the deep sea—specifically the cold, high-pressure benthic zones—will likely render these biological solutions ineffective for the majority of the planet's plastic mass.

Furthermore, I predict the emergence of synthetic triage laws. As the link between the plastisphere and antibiotic-resistant outbreaks becomes clinically undeniable, international maritime law will be rewritten. Nations will be held legally and financially accountable for the microbial export of pathogens hosted on their plastic waste. The future of global health depends on our ability to sever the link between the indestructible polymer and the evolving pathogen, or else face an era where the natural world is permanently replaced by a unified, synthetic-biological hybrid metabolism.

The ultimate terminal fate of the Plasticene is not the disappearance of the polymer, but its total, sub-micron integration into the fabric of life. We are no longer cleaning the environment; we are attempting to manage the internal redesign of the human species.

The Enzymatic Insurgency: Molecular Tactics for the Deconstruction of Synthetic Matter

While mechanical extraction serves as a temporary palliative for macroscopic debris, the true resolution of the polymer crisis lies at the molecular scale. We are currently witnessing a biological insurgency, where the scientific community is attempting to weaponize the catabolic potential of the microbial world to dismantle the synthetic structures we have built. This is not merely an environmental effort; it is a high-stakes engineering challenge focused on identifying and optimizing the "biochemical scissors"—extracellular enzymes—capable of severing the theoretically indestructible bonds of the petrochemical era.

The Crystalline Fortress: Thermodynamic Barriers to Dissolution

The primary obstacle to biological remediation is the internal architecture of the polymer itself. Commercial plastics are not uniform masses; they are composed of a dual-phase geometry consisting of highly ordered crystalline regions and chaotic amorphous zones. The crystalline fortress of materials like high-density polyethylene, which can reach 95% crystallinity, presents an impenetrable physical wall to microbial life.

These tightly packed molecular chains provide no space for enzymes to penetrate, rendering the material effectively invisible to the biological world.

Conversely, the amorphous regions—the "soft" underbelly of the plastic—provide the necessary flexibility and chemical access for enzymatic attack. The rate of degradation is therefore a function of the amorphous-to-crystalline ratio. Most standard petrochemical plastics, including PP, PS, and PVC, are characterized by a carbon-carbon backbone that lacks chemical "weak points." In contrast, polymers featuring heteroatoms like oxygen or nitrogen—such as PET and polyurethane—provide a clear target for enzymatic cleavage. This chemical vulnerability is why PUR and PET are the primary focus of current bio-remediation efforts, while the carbon-carbon recalcitrants remain the ultimate frontier.

The Microbial Strike Force: Consortia and Synergistic Catabolism

Effective remediation in the wild is rarely the work of a single organism. Instead, it requires a microbial consortium—a coordinated strike force of bacteria, fungi, and microalgae working in biological tandem. In these systems, one species may perform the initial "mechanical" etching or surface oxidation, while another secretes the specific enzymes required to cleave the chain into smaller, bioavailable monomers.

Research into extremophiles has revealed that life in the most hostile environments—the freezing Arctic soils and the lightless benthic zones—has already begun to adapt. These organisms possess the latent genetic capacity to treat synthetic matter as a carbon source. For instance, the synergy between specific bacterial genera and fungal mycelial networks has demonstrated the ability to completely mineralize complex polyesters like PCL within a 30-day window under optimized conditions. However, the spatial heterogeneity of these communities means that a solution that works in a warm estuary will be thermodynamically void in the cryogenic sinks of the deep ocean.

Deep Analysis: The Geometry of Colonization A profound, yet overlooked, variable in this process is the geometric influence on taxonomy. Advanced genetic sequencing has confirmed that the specific communities of bacteria and fungi that colonize a plastic particle are dictated by its physical shape. A jagged, irregular shard provides a vastly different topographical niche than a smooth, spherical microbead. These physical variations influence the eco-corona formation and the subsequent depth of enzymatic penetration. Furthermore, weathered plastics—those already scarred by ultraviolet bombardment—exhibit a significantly higher degree of microbial complexity than newly fragmented shards, suggesting that environmental "pre-aging" is a prerequisite for biological integration.

Practical Advice for Remediation Engineering

To bridge the gap between laboratory success and environmental reality, we must shift our strategy from "ocean cleaning" to point-source bio-refining:

- **Thermal Pre-Conditioning:** Highly crystalline plastics like PE must be subjected to "photonic bombardment" (high-intensity UV-C shocks) or localized thermal heating before biological introduction. This artificially shatters the crystalline bonds, expanding the amorphous regions and creating a "foothold" for bacterial attachment.

- **Bioreactor Integration:** Global funding should be redirected from ocean skimming toward the installation of high-temperature enzymatic reactors at municipal wastewater treatment plants. By maintaining an artificial temperature of 45C to 60C, we can maximize the kinetic energy of the enzymes, achieving mineralization rates that are physically impossible in the open sea.

- **The PHA Transition:** We must accelerate the transition to polyhydroxyalkanoates (PHA). These biologically derived polyesters are the only materials currently capable of total mineralization in the marine environment, offering a path to absolute biocompatibility.

Predictions: The Rise of the 'Synthetic Metabolism'

I predict that within the next fifteen years, we will see the emergence of CRISPR-enhanced super-consortia. We will move beyond identifying natural degraders toward "genomic stitching," where the most effective extracellular enzymes from diverse kingdoms are combined into a single, robust microbial chassis. These engineered strains will be designed specifically for the high-pressure, low-temperature environments of the deep sea.

Furthermore, I predict the development of synthetic metabolic pathways. We are approaching a threshold where microbes will not just "break down" plastic, but will utilize the carbon backbone to synthesize high-value biochemicals in situ. The "pollution crisis" will eventually be reframed as a carbon resource opportunity, where the millions of tons of plastic currently choking the oceans are viewed as the raw feedstock for a new, synthetic-biological industrial revolution. The "indestructible" plastic of the 20th century will become the biological fuel of the 22nd, provided we can master the molecular keys to unlock the crystalline fortress.

The Living Scourge: Bio-Fortification and the Microbial Reclamation of the Plastisphere

The fundamental architecture of the bacterial world is defined by a capacity for absolute endurance. These organisms are not merely survivors; they are the architects of a new, synthetic metabolism. Capable of maintaining deep physiological functions within radioactive clouds, sub-zero glacial cores, and high-pressure benthic trenches, bacteria represent the microbial vanguard of the 21st century. When a synthetic polymer enters the marine environment, it is not an inert object; it is an uncolonized territory. Within minutes, a high-density bacterial occupation begins, initiating a process of synthetic bio-fortification that transforms a piece of waste into a thriving, mobile ecological hub.

The Architecture of Occupation: Biofilms and Genetic Warfare

The primary tactic for bacterial survival on a polymer substrate is the formation of the extracellular polymeric substance (EPS). This thick, protective microfilm acts as a biological fortress, shielding the colony from the chaotic chemistry of the ocean while concentrating nutrients at the surface of the plastic. Within these multi-species consortia, a more terrifying phenomenon occurs: horizontal gene transfer (HGT).

Deep Analysis: The Evolutionary Acceleration The plastisphere functions as a high-speed evolutionary laboratory. Because disparate species are packed into the dense EPS matrix, the exchange of genetic material is rampant. This facilitates the rapid emergence of synthetic super-bugs—strains that have evolved the specific genetic instructions to treat polyethylene or polystyrene not as a pollutant, but as a primary caloric fuel. This is not a slow, geological process; it is an aggressive, real-time adaptation that is fundamentally rewriting the microbial map of the oceans.

The Biochemical Siege: Enzymatic Cleavage and the Crystalline Barrier

The true "war" against the polymer is waged through enzymatic secretion. Bacteria release a specialized arsenal of hydrolytic proteins—specifically lipases, proteases, and cutinases—designed to slice through the synthetic carbon-carbon and carbon-heteroatom bonds.

Materials like polyurethane (PUR) are particularly vulnerable because they possess urethane linkages that biologically mimic natural peptide bonds. In a case of "biochemical mistake," bacterial enzymes lock onto these synthetic chains and dismantle them. However, the crystalline barrier remains the greatest obstacle. Highly ordered polymers like high-density polyethylene (HDPE) are effectively impenetrable to enzymes in their raw state. The tight molecular packing leaves no "foothold" for the bacteria to begin their work.

Practical Advice for Remediation Engineering

- **Structural Scoring:** To facilitate microbial attack, industrial-scale remediation must prioritize aero-thermal pre-treatment. Exposing microplastics to high-intensity UV-C radiation or thermal "shocks" creates micro-cracks and expands the amorphous regions of the polymer. This "molecular softening" is a non-negotiable prerequisite; without it, biological degradation in cold water remains a statistical zero.

- **Point-Source Bioreactors:** We must abandon the fantasy of cleaning the open ocean with bacteria. Instead, capital should be redirected toward high-kinetic bioreactors installed at wastewater gateways. By maintaining a constant 45C and optimal nutrient ratios, we can achieve mineralization rates that are 10,000 times faster than those found in the natural environment.

The Extremophile Arsenal: Cryogenic and Abyssal Specialists

As we look to the "cold sinks" of the planet—the Arctic and the deep-sea trenches—we find an elite class of psychrophilic specialists. These cold-adapted bacteria, isolated from ancient glaciers and the Mariana Trench, have evolved enzymes that remain active even at temperatures hovering near 0C.

Specific strains of Pseudomonas and Bacillus have demonstrated the capacity to produce active lipases that function under crushing hydrostatic pressure. This "abyssal labor force" represents the only viable long-term strategy for addressing the legacy plastic currently settling into the lightless benthic graveyards. While their natural rates are slow, these strains provide the genetic "blueprint" for future industrial enzymes.

Predictions: The Shift to Synthetic Metabolic Sovereignty

I predict that within the next decade, we will witness the end of natural reliance. Remediation firms will no longer wait for the ocean to "evolve" a solution. Instead, we will see the rise of custom bio-consortia. Utilizing CRISPR-Cas9 technologies, researchers will "stitch" the most effective enzymes from extremophiles into robust, industrial bacterial chassis. These synthetic consortia will be engineered for total annihilation, capable of reducing PET or PVC to CO_2 and H_2O in a matter of hours.

Furthermore, I predict the emergence of landfill injection systems. As our understanding of these "super-strains" matures, they will be forcefully injected into the millions of tons of buried synthetic waste in municipal landfills. We will move from a strategy of storage to one of "accelerated biological collapse," turning the plastic graveyards of the 20th century into the carbon-rich biological fields of the 21st.

The Mycelial Siege: Fungal Biomechanical Warfare and the Erosion of the Synthetic Frontier

While the bacterial world occupies the liquid water column, the true "executioners" of the synthetic age are found within the fungal kingdom. Fungi possess a unique, dual-mode evolutionary advantage over their bacterial counterparts: they do not merely wait for chemical diffusion; they engage in active mechanical warfare. The expansion of mycelial networks acts as a hydraulic drill, physically penetrating the crystalline lattice of the plastic matrix. This structural shattering creates a massive, fresh surface area that facilitates a secondary, hyper-localized enzymatic assault. By forcing the insertion of oxygen-rich functional groups—specifically carbonyl (=O), carboxyl (-COOH), and ester bonds—into the inert carbon-carbon backbone, fungi strip away the polymer's inherent hydrophobicity, effectively "priming" the plastic for total biological dissolution.

The Polyurethane Executioners: Specialized Fungal Metabolism

Forensic audits of environmental degradation reveal that fungi are the primary biological agents responsible for the deconstruction of complex polyester polyurethane (PUR). While bacteria often struggle with the structural complexity of these industrial foams, specific fungal lineages, particularly within the Fusarium and Rhizopus genera, have evolved highly concentrated secretomes. These organisms produce specialized lipases and esterases that target the urethane linkages—chemical structures that deceptively mimic the peptide bonds of natural proteins.

This biochemical mimicry allows fungi to "recognize" PUR and PET as potential energy sources. Furthermore, the fungal capacity for remediation is not limited by environmental acidity or alkalinity. Analytical data from contaminated industrial sites confirm that species such as Geomyces pannorum and Phoma function as dominant degraders regardless of extreme pH fluctuations.

The Arctic and Marine Frontiers: Extremophile Mycology

The search for biological "super-strains" has moved to the planet's most hostile extremes. From the sub-zero soils of the Arctic to the dynamic coastal zones of the Indian Ocean, specialized "psychrotolerant" (cold-adapted) fungi have been isolated. Strains of Clonostachys rosea and Trichoderma demonstrate a remarkable capacity to maintain metabolic activity under severe thermal stress, while marine-adapted Aspergillus species have achieved the "impossible": the degradation of high-density polyethylene (HDPE).

Deep Analysis: The Hyphal Advantage in the 'Dry Sink' It is a critical error to overlook the "dry sink" of terrestrial microplastics. While much of the global focus remains on the oceans, agricultural topsoils are currently being choked by PVC irrigation pipes and polyethylene mulch films. Bacteria are largely ineffective in these environments due to their reliance on high moisture levels. Fungi, however, thrive in the low-moisture, aerated environment of the topsoil. Their mycelial threads act as a "synthetic information network," moving nutrients and enzymes across vast distances through the soil matrix to reach isolated plastic shards.

Practical Advice for Soil and Industrial Remediation

- **The Topsoil Mandate:** When treating agricultural lands contaminated with mulch films, mycoremediation must be prioritized over bacterial treatments. Fungi are better equipped to handle the mechanical resistance of buried plastic and the fluctuating moisture levels of the soil surface.

- **Bioreactor Scaling:** Industrial-scale bio-refineries should utilize solid-state fermentation (SSF) rather than liquid-batch processing. By growing Penicillium and Aspergillus on a substrate of shredded PET or PU waste, we can stimulate the massive production of "plastic-targeted lipases."

- **Pre-Colonization:** For unavoidable industrial plastics, such as marine aquaculture nets, the materials should be "pre-colonized" with dormant fungal spores that activate upon exposure to seawater, initiating a self-destruct sequence the moment the plastic begins to wear.

Predictions: The Rise of the Fungal Bio-Foundry

I predict that within the next fifteen years, we will see the emergence of fungal bio-foundries. These will be massive, automated facilities where synthetic consortia—genetically optimized combinations of Trichoderma and Rhodococcus—will be used to transform plastic waste into biogenic fuels and building materials.

Furthermore, I predict the development of mycelial scaffolding for marine restoration. We will utilize plastic-degrading fungi to break down existing "ghost nets" in the ocean, while simultaneously using the fungal biomass to provide a nutrient-rich substrate for coral reef regrowth. The synthetic legacy of the 20th century will eventually be "digested" into the biological record, serving as the raw feedstock for a planet-wide mycelial revolution.

The Mycelial Infiltration: Advanced Fungal Processing and the Mechanics of Synthetic Dissolution

The global dominance of polyethylene (PE) is a testament to human engineering and a curse to biological cycles. Because PE is designed for absolute durability and structural indifference to the environment, its remediation requires more than just passive bacterial contact. It requires an active, aggressive biological invasion. Fungi, through their unique mycelial architecture, offer a mechanical advantage that bacteria cannot match. While bacterial colonies form thin, vulnerable films, fungal networks act as biological "drills," physically penetrating the smooth, hydrophobic surfaces of the polymer and creating a high-surface-area bridge between the synthetic carbon and the aqueous world.

Specialized Biological Assets: Marine and Gut-Derived Catalysts

The search for effective polymer degraders has led to the identification of specialized "polymer executioners." In the nutrient-starved, high-salinity environments of the open ocean, the fungus Zalerion maritimum has emerged as a critical asset. It does not merely live on plastic; it utilizes the carbon backbone of PE as a primary metabolic fuel, effectively shrinking the mass and volume of synthetic debris without the need for external nutrient supplementation.

Even more promising is the "gut-derived infiltration" strategy. By mining the microbiomes of organisms evolved to consume wax—such as the larvae of the Galleria mellonella—researchers have isolated hyper-aggressive strains of Aspergillus flavus. These gut-derived catalysts produce specialized laccase-like multicopper oxidases (LMCOs) that can reduce the molecular weight of high-density polyethylene (HDPE) within weeks. This process is marked by the violent grafting of oxygen-rich functional groups—specifically carbonyl and ether moieties—directly onto the polymer chain, effectively "opening" the material for final enzymatic destruction.

The Ecology of the 'Synthetic Reef'

It is a profound ecological error to assume that the fungi colonizing plastic are the same as those colonizing natural wood or stone. We are seeing the birth of a distinct "plastisphere taxonomy." Fungal communities found on PE and polystyrene (PS) are genetically and structurally distinct from their natural counterparts. These "synthetic assemblages" are highly sensitive to geography and environment; a fungal mat in the Baltic Sea possesses a fundamentally different genetic toolkit than one found in a municipal wastewater plant. The dominance of saprophytic phylums like Ascomycota and Chytridiomycota in these environments suggests that fungi are rapidly evolving to recognize synthetic polymers as a new, primary ecological niche.

The Biochemical Engine: Depolymerization and Mineralization

To dismantle synthetic chemistry, biology must perform a coordinated metabolic heist. Heterotrophic microorganisms act as the planet's deconstructive engines, utilizing a sequence of biochemical reactions to convert complex polymers back into their foundational elements. This process, known as mineralization, is the terminal stage of the waste cycle, where the plastic is finally reduced to CO_2 and H_2O.

The work is done by exo-enzymes—biological catalysts secreted into the immediate surroundings. These proteins perform depolymerization through two primary modes: aggressive oxidation and targeted hydrolytic cleavage. This attack results in a measurable decline in the plastic's molecular weight and a total collapse of its tensile strength. However, this is an energy-intensive process; the microbes are not "cleaning" the environment by choice, but are ruthlessly extracting the stored chemical energy within the polymer bonds to fuel their own cellular reproduction.

The Physics of Resistance: The Crystallinity Barrier

The speed of this biological heist is dictated by the polymer's internal physics. Synthetic plastics are composed of two distinct regions: the ordered crystalline zones and the chaotic amorphous zones. Because H_2O molecules and bulky enzymes cannot penetrate the dense, impenetrable crystalline lattice, the initial attack is always focused on the amorphous regions. Polymers with high crystallinity, like commercial PE (often exceeding 95% crystallinity), remain functionally indestructible in the wild because their molecular architecture is too tightly packed for biological "tools" to find a foothold.

Practical Advice for Remediation Engineering

- **Thermodynamic Softening:** When designing industrial-scale bioreactors for municipal waste, the temperature must be hyper-managed. By heating the microplastic slurry to a point just below the polymer's Glass Transition Phase, we can artificially expand the amorphous regions. This "thermal softening" increases the physical penetration rate of the introduced LMCOs and exo-enzymes, accelerating the rate of mineralization by orders of magnitude compared to ambient field conditions.

- **Surface Scouring:** Prior to biological inoculation, PE waste should be mechanically scoured or chemically etched to create artificial surface textures, providing a "grip" for mycelial attachment.

Predictions: The Deployment of Fungal 'Bio-Fences'

I predict that the future of riverine and coastal protection will shift away from mechanical skimming and toward the deployment of fungal bio-fences. We will see the mass deployment of pre-seeded, floating mycelial "mats"—biodegradable networks of engineered fungi designed to physically trap floating PE and PS debris in river mouths. These mats will act as living filters, initiating the enzymatic deconstruction of the plastic before it can ever reach the open ocean.

Furthermore, I predict a "genetic convergence" where the PETase and LMCO sequences from various extremophiles are spliced into a single, robust fungal host. This will create a universal degrader—a fungal organism capable of dismantling PET, PE, and PVC simultaneously. The synthetic legacy of the 20th century will eventually be "digested" into the biological record, serving as the raw feedstock for a planet-wide mycelial revolution where waste is once again a biological currency.

The Thermodynamic Descent: Bio-Kinetic Pathways and the Tripartite Architecture of Synthetic Erasure

The destruction of a synthetic polymer is not a disappearance; it is a thermodynamic descent from a high-energy, complex solid to a low-energy, elemental gas. In the natural environment, this transition is governed by the laws of chemical kinetics and the metabolic efficiency of the microbial world. To understand the terminal fate of the "Plasticene," we must analyze the polymer as a carbon-dense fuel source and the microorganisms as biological reactors. Whether through the rapid, oxidative efficiency of the surface waters or the slow, methane-heavy decay of the benthic sludge, the ultimate goal of the biosphere is the total mineralization of the synthetic intruder.

The Kinetic Supremacy of Aerobic Metabolism

The efficiency of polymer breakdown is dictated by the availability of the terminal electron acceptor. Under highly oxygenated, aerobic conditions, the biological processing of a carbon-based polymer follows a high-yield stoichiometric pathway. In this state, molecular oxygen (O_2) facilitates the most energetically favorable respiratory chain. The biological extraction of energy from the carbon backbone is at its peak, allowing microbial communities to proliferate with an oxidative advantage. This rapid growth enables the colony to aggressively colonize the plastic substrate, turning the polymer's stored energy into fresh cellular biomass while exhausting CO2 and H2O as metabolic byproducts.

Conversely, when the environment shifts to an anaerobic (oxygen-depleted) state—such as the interior of a landfill or the anoxic "dead zones" of the sea—the thermodynamics collapse. Deprived of oxygen, microbes must rely on less efficient electron acceptors like nitrates, sulfates, or iron. The metabolic exhaust shifts to include CH4 (methane), and the kinetic rate of degradation plummets. This anaerobic lag is why plastic buried in deep sediments remains functionally permanent; the energy yield is simply too low to sustain aggressive biological dismantling.

The Tripartite Architecture of Deconstruction

The transition from a solid plastic bottle to atmospheric gas is a sequential, three-stage "inquest" of the material. Each stage requires a different set of tools, moving from the physical violence of the environment to the surgical precision of the enzyme.

Stage 1: Surface Erosion (Biodeterioration) The first phase is a "softening of the defenses." Before a microbe can ingest the polymer, the material must undergo biodeterioration—a catastrophic loss of its mechanical integrity. This is driven by abiotic violence: ultraviolet bombardment, thermal expansion, and wave-induced friction. These forces create a "micro-topography" on the plastic surface, increasing its hydrophilicity and providing the necessary "grip" for microbial attachment. Paradoxically, the high-density adsorption of toxic pollutants onto the plastic may actually accelerate this stage by providing a localized nutrient "bounty" that attracts the primary colonizers.

Stage 2: Enzymatic Dismantling (Biofragmentation) Once the surface is compromised, the microbial community deploys its "biochemical sledgehammers"—extracellular enzymes known as exo-enzymes. Through a process of biofragmentation or depolymerization, enzymes like lipase, esterase, and proteinase K target the polymer chains. They cleave the long carbon-carbon or heteroatom bonds, reducing the massive molecular weight of the plastic into a slurry of short-chain oligomers, dimers, and finally, individual monomers. This is a targeted strike: the enzymes focus on the "amorphous zones" where the molecular packing is loose, bypassing the "crystalline fortress" of the polymer until the very end.

Stage 3: Metabolic Hijacking (Assimilation) The final stage is the total internalization of the synthetic carbon. In the assimilation phase, the newly liberated monomers are identified by specific cellular receptors and violently absorbed across the microbial cell membrane. Once inside the intracellular matrix, the plastic carbon is no longer "waste"—it is the raw feedstock for the synthesis of ATP, the energy currency of life. Through complex biochemical loops, the synthetic carbon is converted into cellular packaging, fresh biomass, and secondary metabolites. True mineralization is achieved only when the last carbon atom is exhausted as gas (CO_2 or CH_4), effectively erasing the synthetic record from the ecosystem.

Deep Analysis: The Residual Dust and Thermal Feedback A critical oversight in contemporary remediation is the "recalcitrant fraction." A mass balance reveals that 100% of the plastic is almost never converted into gas or biomass in the wild. A stubborn percentage of highly degraded, ultra-concentrated carbon dust often remains, which may be more toxic than the original polymer. Furthermore, we must account for the exothermic feedback: as microbes dismantle the plastic, they release thermal energy. In high-density "plastisphere" colonies, this localized heat may further soften the polymer, creating a self-reinforcing loop of degradation that bypasses ambient temperature limitations.

Practical Advice for Industrial Bioremediation

- **The Aerobic Surge:** In municipal bioreactors, the "anaerobic switch" must be avoided at all costs. Continuous oxygen saturation is required to maintain the kinetic supremacy of the degradation path.

- **Pre-Fragility Engineering:** We must prioritize the manufacture of "pro-oxidant" polymers that are designed for rapid Stage 1 biodeterioration. If the material cannot shatter into bioavailable fragments within a specific window, Stage 3 mineralization will never be reached.

- **Enzymatic Priming:** Remediation efforts should utilize a "pre-enzyme soak" using broad-spectrum proteases like proteinase K to "score" the material before microbial inoculation, essentially bypassing the slow natural biodeterioration phase.

Predictions: The Advent of 'Directed Degradation'

I predict that within the next fifteen years, we will transition from "passive observation" to directed degradation. We will no longer rely on the random survival of wild bacteria. Instead, we will deploy synthetic metabolic cascades—engineered microbial consortia where each species is optimized for one specific stage of the tripartite model. One strain will handle the physical erosion, the second will cleave the chains, and the third will perform the final intracellular mineralization.

Furthermore, I predict the emergence of carbon-trace treaties. As the ability to track the C-biomass of plastic-eating microbes improves, we will see international laws that demand the "total molecular erasure" of all produced polymers. The "Plasticene" will end not when we stop making plastic, but when we master the thermodynamic keys to turn it into thin air the moment its utility expires.

The Molecular Guillotine: Engineering the Enzymatic Siege of Synthetic Carbon

The terminal deconstruction of a synthetic polymer is not a result of biological chance, but of a highly specific "molecular siege" conducted by a specialized elite of the microbial world. While over 90 distinct taxa have been identified as having the genetic blueprint for polymer catabolism, the transition from an inert industrial solid to a bio-available nutrient is a complex, multi-stage biochemical heist. This process is governed by the absolute specificity of extracellular enzymes—biological catalysts that must operate in the chaotic environment of the open sea or the nutrient-dense silt of a landfill.

The Hydrophilic Pivot: Priming the Surface for Destruction

The primary challenge of any polymer-degrading enzyme is the material's inherent hydrophobicity. To dismantle a synthetic chain, the biological world must first force a "chemical pivot." This is achieved through hydrolysis, the critical first step in the mineralization sequence. During this phase, specialized hydrolase enzymes—including glycosidases, lipases, and phosphatases—violently graft oxygen-rich functional groups directly onto the inert polymer backbone.

By introducing highly reactive single-bonded (C-O) and double-bonded (C=O) carbonyl and ether moieties, the enzyme artificially increases the surface's affinity for water. This oxidative "priming" softens the material's defenses, allowing water-soluble enzymes to penetrate the matrix.

Surgical vs. Structural Cleavage: The Mechanics of Exo and Endo Attacks

The deconstruction of the polymer chain follows two distinct tactical pathways:

- **The Surgical Exo-Attack:** In this configuration, the enzyme acts as a molecular "peeler," sequentially cleaving terminal monomers from the ends of the long-chain polymer. These singular, low-molecular-weight pieces are highly soluble and can be instantly absorbed across the microbial cell membrane for energy.

- **The Structural Endo-Attack:** This involves the random "slashing" of the polymer backbone at internal weak points. While this rapidly shatters the macro-structure into smaller fragments (oligomers), these pieces often remain too massive for immediate cellular ingestion.

The Cross-Kingdom Arsenal: Bacterial and Fungal Toolkits

- **The Bacterial Toolkit:** Specialized marine bacteria produce robust PUR esterases, alkene monooxygenases, and PEG laccases designed to target the unique molecular linkages of industrial foams and rigid plastics.

- **The Fungal Toolkit:** Fungal species, with their penetrating mycelial architecture, deploy an even more aggressive array, including manganese peroxidases, cutinases for shredding PCL, and broad-spectrum proteases.

Practical Advice for Bioreactor Engineering

To move from laboratory theory to industrial-scale remediation, engineers must embrace enzymatic synergy:

- **The Cocktail Strategy:** Never rely on a single-enzyme system. Rapid deconstruction requires a "synergistic cocktail" that combines high-torque endo-enzymes (to maximize exposed surface area via fragmentation) with high-efficiency exo-enzymes (to ensure immediate mineralization of the resulting monomers).

- **Amorphous Accessibility:** To overcome the crystalline barrier, pretreatment protocols must utilize a "thermal pulse" to shift the polymer toward its glass transition phase, artificially expanding the amorphous "attack zones" before the introduction of the enzymatic cocktail.

Predictions: The Advent of Recombinant Enzyme Therapy

I predict that within the next decade, the reliance on living, "temperamental" microorganisms for environmental cleanup will be completely superseded by cell-free recombinant therapy. Utilizing AI-driven protein folding models, biotechnology firms will design "super-enzymes" that combine the mechanical drilling power of fungal peroxidases with the high-speed monomer peeling of bacterial hydrolases.

These synthetic proteins will be mass-produced in industrial vats and stabilized for environmental deployment. We will see the rise of enzymatic aerial seeding, where stabilized, dry-form enzymes are sprayed over massive landfill sites or coastal "hot zones." These enzymes will operate as cell-free chemical agents, initiating the rapid collapse of synthetic waste without the need to maintain a living microbial culture in a toxic environment. The Plasticene will finally be dismantled not by the slow march of evolution, but by the high-speed execution of engineered biochemistry.

The Synthetic Reef: Bio-Architectural Colonization and the Inversion of Buoyancy

The introduction of an inert, synthetic polymer into the marine environment is not a passive event; it is the immediate creation of a high-stakes ecological frontier. Within the first temporal window of exposure, the naked landscape of the plastic is targeted by an "enzymatic handshake." Opportunistic microorganisms, drifting in a chaotic planktonic state, undergo a radical genetic re-programming upon contact with the polymer. They transition from free-floating individuals into a highly structured, sessile community, constructing what is effectively an "organo-synthetic pellicle" or biofilm. This colonization is driven by the secretion of polymeric secretory matrices (PSM)—a sticky, molecular fortress of lipids, signaling proteins, and extracellular DNA that cements the biological world to the industrial one.

The Phases of Molecular Occupation

The occupation of the plastic surface follows a rigid, genetically mandated tripartite cycle:

1. **Irreversible Adhesion:** The immediate transition where cells flip their genetic switches to produce adhesives that anchor them against hydraulic shear.

2. **Structural Maturation:** The development of a three-dimensional architecture where the colony optimizes nutrient flow and protective layering.

3. **Strategic Dispersal:** Once the localized synthetic or organic resources are exhausted, the community "abandons ship," reverting to a planktonic state to find fresh substrata.

Deep Analysis: The Nutrient Paradox The relationship between the biofilm and its synthetic host is complex. Adsorbed organic pollutants can act as a biological bounty, providing the energy required for rapid colony expansion. Conversely, the "chemical hemorrhaging" of plasticizers (like phthalates) can act as a localized toxin, triggering cellular apoptosis and preventing colonization. Thus, the biofilm is a living sensor, reflecting the precise chemical toxicity of the underlying polymer.

The Buoyancy Inversion: Life as a Biological Anchor

One of the most profound hydrodynamic consequences of this colonization is the inversion of buoyancy. Synthetic polymers like PE and PP are engineered to be lightweight, typically floating at the surface. However, the rapid accumulation of biological mass—from pioneer bacteria to secondary layers of diatoms, algae, and eventually macro-crustaceans like barnacles—radically increases the specific density of the particle.

This process, termed bio-anchoring, forces the plastic to lose its buoyancy and sink through the stratified water column. The plastisphere thus acts as a vertical shuttle, carrying surface-level pollutants down into the lightless benthic zones. As the microbes "bore" into the polymer using highly acidic metabolic byproducts, they increase the surface-to-volume ratio, further accelerating the degradation and the rate of descent.

The Vector Threat: Synthetic Hitchhiking

The plastisphere represents a novel, artificial biome that facilitates synthetic hitchhiking. Pathogenic or invasive species, protected within the PSM fortress, are transported across vast distances by ocean currents. This allows for the biological contamination of pristine ecosystems that were previously isolated by geographical distance. We are witnessing a borderless dispersal of alien genetic material, powered by the indestructible nature of the polymer carrier.

Practical Advice for Industrial Bioremediation

- **Starvation-Induced Adhesion:** For industrial bio-filters, the bacterial culture (specifically robust strains like Rhodococcus) should be subjected to severe carbon starvation before being introduced to the plastic. In a state of biological desperation, these bacteria develop a hyper-hydrophobic cell surface, allowing them to anchor to the slick plastic surface with exponentially greater force.

- **Surface Topography Engineering:** Pre-treating waste streams with localized UV shocks creates a "pitted" landscape. This increased surface roughness provides the mechanical "grip" required for the primary bacterial pioneers to establish a stable foundation.

Predictions: The Advent of Autonomous Bio-Scrubbers

I predict that within the next decade, municipal wastewater infrastructure will undergo a labyrinthine retrofit. We will move away from static chemical treatments toward the installation of autonomous bio-scrubbers. These will be massive, high-surface-area matrices of biodegradable polymers pre-colonized with hyper-hydrophobic, starved biofilms.

These living filters will be installed at the final effluent gates of every urban treatment plant. They will act as selective molecular traps designed to strip, anchor, and enzymatically digest nanoscale synthetic fibers from laundry runoff before they can enter the open sea. Furthermore, I predict the emergence of satellite-linked bio-buoys— autonomous remediation units that track plastic patches via satellite and deploy microbial strike teams to induce buoyancy inversion, forcing the plastic into localized deep-sea collection zones.

The Kinetic Ceiling: Stoichiometric Barriers and the Molecular Metrics of Synthetic Decay

The transition from a stable industrial polymer to a mineralized elemental state is not a guaranteed biological trajectory; it is an uphill thermodynamic struggle governed by a rigid set of bottlenecks. These impediments exist in two distinct dimensions: the environmental volatility of the surrounding habitat and the intrinsic molecular architecture of the polymer itself. To achieve true bioremediation, we must navigate the complex interplay between these domains, where a shift in a single variable—such as a 1°C drop in temperature or a marginal increase in molecular crystallinity—can effectively halt the catabolic process, leaving the plastic in a state of permanent environmental suspension.

The Metrics of Failure: Beyond Gravimetric Analysis

Historically, the success of plastic degradation has been measured through the crude metric of "weight loss." However, a deep analytical audit reveals this to be a profoundly deceptive indicator. In many cases, a recorded loss in mass does not represent the removal of carbon from the ecosystem, but rather the catastrophic fragmentation of the material into invisible, sub-micron nanoplastics. This mass-to-particle transition creates an illusion of remediation while actually escalating the ecological risk.

To establish a valid forensic baseline, the scientific community must shift toward a hierarchy of molecular metrics:

- **Stoichiometric Evolution:** The absolute, mathematical quantification of CO_2 gas evolution via closed-system respirometry. This is the only definitive proof that the synthetic carbon backbone has been fully mineralized and recycled into the atmosphere.

- **Structural Fragility:** Measuring the catastrophic decline in mechanical tensile strength and elasticity. This provides a window into the internal rupturing of the material, where water uptake and enzymatic penetration cause the polymer to swell and fail from the inside out.

- **Vibrational Fingerprinting:** Utilizing FTIR spectroscopic shifts to track the formation of reactive oxygen species and the introduction of hydrophilic functional groups, marking the biological softening of the plastic face.

The Physicochemical Architecture of Resistance

The intrinsic bottlenecks are rooted in the polymer's crystalline-to-amorphous ratio. A polymer like high-density polyethylene is effectively a molecular fortress; its chains are so tightly packed that neither H_2O molecules nor bulky biological exo-enzymes can find a foothold. Degradation only begins when environmental stressors—such as ultraviolet bombardment—induce enough structural chaos to create amorphous pockets. Without this prerequisite abiotic priming, the microbial community is physically locked out of the carbon source, regardless of its genetic potential.

Deep Analysis: The Swelling Threshold One of the most significant, yet overlooked, indicators of successful microbial infiltration is the water uptake capacity. As enzymes begin to cleave the polymer chains, the density of the material decreases, allowing water to penetrate the matrix. This leads to a localized, internal expansion that physically shatters the crystalline lattice. This internal mechanical pressure is a more reliable predictor of terminal mineralization than any surface-level observation.

Practical Advice for Environmental Forensic Teams

- **The Respirometry Mandate:** Researchers must move beyond simple weight-loss studies. Verifiable degradation must be strictly measured by capturing the exact molar volume of CO_2 produced. If you cannot account for the carbon gas, you cannot claim the plastic is gone.

- **Tensile Sensitivity:** In field studies, prioritize the measurement of tensile loss as an early-warning indicator of biological colonization. A drop in mechanical strength often precedes any measurable loss in mass and provides a more accurate view of the systemic failure of the polymer.

Predictions: The Rise of Real-Time Stoichiometric Monitoring

I predict that within the next decade, we will see the emergence of in-situ stoichiometric sensors. We will move away from laboratory batch testing toward the deployment of autonomous buoys equipped with carbon-isotope trackers. These sensors will be able to distinguish between CO_2 produced by natural biological respiration and CO_2 produced by the mineralization of synthetic C-based polymers.

Furthermore, I predict a global standardization of the 'Mineralization Coefficient'. Regulatory bodies will eventually demand that every commercial plastic product be assigned a degradation half-life based on its molecular architecture. This will lead to a new era of programmable decay, where the rate of a polymer's return to the elemental state is as carefully engineered as its initial durability.

The Synthetic Resurrection: Engineering the Molecular Rebirth of Industrial Waste

The geopolitical landscape of the mid-21st century is defined by a fundamental collision between industrial acceleration and ecological solvency. While the global demand for polymers continues to grow, we are standing at the precipice of a molecular realignment. The crisis of pervasive microplastic infiltration has forced a shift in biological research: we are no longer looking at plastic as an indestructible waste product, but as a sequestered carbon resource waiting for a key. By moving away from passive observation toward a state of engineered synergy, we can begin the high-speed deconstruction of the petrochemical era.

The Extremophile Library: Mining the Genomic Frontier

The scientific community has, until now, only explored the microscopic crust of the planet's microbial potential. The true catalysts for a post-polymer world are hidden in the Extremophile Library—the unmapped genetic diversity thriving in the planet's most hostile niches. From the crushing pressures of the abyssopelagic zones to the sub-zero cryospheres of the poles, life has already begun to solve the petrochemical problem.

Deep Analysis: The Niche-to-Lab Pipeline These extreme environments—anoxic wetlands, toxic municipal sewer lines, and hyper-compacted landfills—are effectively evolutionary pressure cookers. The bacteria and fungi thriving in these zones have already developed the metabolic pathways to treat synthetic carbon as a primary energy vector. The challenge of the next decade is not merely to find these organisms, but to map their omics signatures—utilizing metagenomics and proteomics to identify the exact biochemical sequences responsible for bond-cleavage.

Entomological Reactors: The Macro-Biological Vector

While microbial research dominates the discourse, the Entomological Reactor offers a unique, high-throughput solution. Groundbreaking discoveries in the gut microbiomes of specific larvae—notably the wax worm (*Galleria mellonella*) and the mealworm (*Tenebrio molitor*)—have revealed an innate capacity to physically masticate and chemically degrade polyethylene and polystyrene.

Rather than viewing these as pests, future remediation strategies will utilize them as primary biological processors. We are moving toward a state where entomological biorefineries will be used to process bulk plastic waste. By culturing the specific endoenzymes discovered within these larval tracts, we can create a scalable, macro-biological infrastructure that shatters the polymer matrix before it ever enters the oceanic sink.

The CRISPR Revolution: Weaponizing the Microbial Secretome

The most disruptive shift in remediation will be the transition from discovery to design. Utilizing CRISPR-Cas9 platforms, we can now rewrite the microbial secretome. We are no longer limited by the slow pace of natural evolution. Instead, we can stitch together the most aggressive enzymes from diverse kingdoms into a single, robust microbial chassis.

Prediction: The Advent of Directed Evolution I predict that within the next fifteen years, we will witness the deployment of targeted super-strains. These will be genetically locked organisms designed to thrive in specific environments with a single metabolic directive: the absolute mineralization of synthetic waste. These strains will be engineered with biological kill-switches that trigger cellular apoptosis once the plastic substrate is exhausted, preventing the risk of environmental runaway.

From Waste to Wealth: The Upcycling Paradigm

The ultimate terminal fate of plastic waste is not its disappearance, but its conversion into value. We are moving away from simple recycling toward molecular upcycling. Through advanced microbial fermentation, the toxic shards of the Plasticene can be enzymatically converted into high-value commodities: alkanes, bio-lubricants, and advanced biofuels.

This shifts the entire economic logic of the crisis. Plastic management will transition from a taxpayer-funded liability into a lucrative global commodity market. The millions of tons of plastic currently choking the oceans will be viewed as the petroleum of the 22nd century, harvested and refined not through drilling, but through biological processing.

Practical Advice for Industrial Synthesis: Design for Decay

To prevent the next generation of environmental collapse, the remediation pipeline must be integrated into the point of manufacturing.

- **Pro-Oxidant Integration:** Industrial synthesis must be legally mandated to include UV-sensitive photosensitizers and chemical pro-oxidants within the virgin plastic matrix. This ensures that once the product enters the waste stream, its molecular "immune system" is already compromised, making it instantly vulnerable to microbial attack.

- **The Biocide Ban:** We must aggressively reduce the use of synthetic antioxidant stabilizers and industrial biocides that are currently added to extend a product's life. These chemicals act as "biological shields" that prevent the very remediation we are trying to achieve.

The Post-Pandemic Imperative: Medical-Grade Bioplastics

The global legacy of the COVID-19 pandemic—a "polypropylene tsunami" of masks and PPE—has created an emergency ecological disaster. The future of medical logistics must be built on truly biodegradable medical-grade bioplastics.

I predict the development of antiviral biopolymers—materials coated with natural, drug-active agents that prevent pathogen proliferation while remaining 100% digestible by soil fungi. This would allow for the direct composting of hospital waste, turning a high-risk biohazard into a nutrient-rich agricultural resource. The era of the "disposable mask" will be replaced by the era of the "dissolvable membrane," where the safety of the patient is no longer bought at the cost of the biosphere.

The "synthetic resurrection" is not a distant hope; it is a technological inevitability. By mastering the molecular keys of the microbial and entomological worlds, we are rewriting the geological record of our planet, ensuring that the legacy of our production is not one of permanent decay, but of biological renewal.

The Point-Source Crisis: The Obsolescence of Legacy Infrastructure

The current global infrastructure for water treatment is suffering from a legacy of biological bias. Most WWTPs were designed decades ago to manage organic human waste and agricultural runoff through rudimentary physical settling and chemical flocculation. While these methods are effective for high-density organic solids, they are essentially transparent to the low-density, non-polar polymers that characterize modern micro-pollution. Consequently, the very facilities intended to protect the environment have become concentrated point-source emitters, funneled by urban drainage and industrial discharge.

To sever this pipeline, we must move toward the tertiary polishing of all municipal effluents. Advanced membranes act as the final, absolute gatekeeper, specifically targeted at the sub-100 µm fraction that survives primary clarifiers. By integrating MBR systems, which combine biological treatment with high-precision physical separation, we can achieve a state of "absolute exclusion," where even sub-micron biological pathogens and synthetic fibers are mechanically stripped from the water column.

Deep Analysis: The Strategy of Downcycling To achieve true sustainability, we must implement a hierarchy of utility for spent membranes. A degraded RO membrane, while no longer suitable for high-pressure desalination, remains a potent tool for secondary brackish water softening or as a coarse pretreatment layer for primary systems. However, the ultimate terminal goal must be the development of bio-sourced filtration matrices—membranes constructed from naturally degradable polymers that can be composted once their mechanical life cycle is complete.

Practical Advice for Municipal and Industrial Retrofitting

- **The MBR Priority:** Municipalities should prioritize the integration of membrane bioreactors over traditional secondary aeration tanks. The MBR provides a smaller footprint and a vastly superior exclusion rate for the microfibers shed from domestic laundry.

- **Decentralized Filtration:** We must move toward point-of-generation defense. Industrial factories, particularly those in the textile sectors, should be legally mandated to install high-pressure ultrafiltration units directly on their internal drainage lines.

- **Thermal Protection:** To prevent the premature "blinding" or clogging of membranes, industrial effluents should undergo a thermal stabilization phase. Cooling the water prior to filtration preserves the sieve geometry.

Steric Blockades and the Polyethylene Density Paradox: Engineering Ultrafiltration Synergies

In the hierarchy of aqueous remediation, ultrafiltration (UF) represents the most energetically efficient steric barrier currently available for high-volume municipal and industrial applications. Operating at a relatively low barotropic pressure of 1 to 10 bar, UF utilizes a precise molecular sieve with a pore hierarchy of 1 to 100 nm. This architecture does not merely filter; it mechanically annihilates the passage of suspended solids, microplastics, and the most resilient biological pathogens, establishing a zero-pass standard for treated effluents.

The Pathogenic Blockade: Beyond Chemical Disinfection

One of the most significant engineering attributes of the UF matrix is its ability to bypass the limitations of traditional chlorination. Highly resilient parasitic organisms, specifically *Giardia* and *Cryptosporidium*, possess a high degree of chlorine resistance that allows them to survive standard secondary and tertiary treatment. UF technology renders this resistance irrelevant by physically blocking their passage. By achieving 99-100% removal of waterborne pathogens and a 95% reduction in biochemical oxygen demand (BOD), UF provides a level of biological and physical security that legacy sedimentation tanks and slow sand filters cannot replicate.

For heavy industry, the implementation of internal, closed-loop UF systems facilitates a transition to zero-liquid discharge (ZLD). In sectors such as polymer manufacturing and commercial paper bleaching, the ability to reclaim and recirculate process water represents not only an ecological victory but a significant shift in operational profitability.

The Buoyancy Failure: Under standard gravitational settling conditions, the removal efficiency for PE microplastics (specifically those < 0.5 mm) is a dismal under 20%. These particles do not sink; they glide through the clarifiers, evading mechanical skimmers and entering the riverine systems. To overcome this density bottleneck, we must move away from passive settling toward chemically weighted flocculation.

Practical Engineering Advice: Optimizing Coagulation Synergies

To maximize the efficiency of an ultrafiltration plant, facility engineers must prioritize the weighted slurry approach:

- **Targeted PAM Dosing:** The addition of specialized, high-molecular-weight polyacrylamide (PAM) to the influent is essential. PAM acts as a molecular glue, binding neutrally buoyant PE fragments into heavy, dense flocs.

- **Micro-Screening Pre-Treatment:** To prevent membrane blindness and catastrophic fouling, high-surface-area micro-screens must be installed prior to the UF stage. These screens capture the aerodynamic slender fibers shed from synthetic textiles, which are notorious for weaving into membrane pores and permanently diminishing flux rates.

- **Flux Stabilization:** In industrial ZLD systems, maintain a constant cross-flow velocity rather than a dead-end filtration dynamic. This prevents the formation of a compressed plastic cake on the membrane surface, extending the interval between chemical cleaning cycles.

Strategic Predictions: The Future of Modular ZLD

I predict that within the next decade, the mega-plant model of wastewater treatment will be superseded by modular, distributed ZLD units. Rather than funneling all urban waste to a central location, high-efficiency UF modules will be installed at the point of generation—specifically in high-density residential complexes and industrial parks. These units will treat and recirculate water locally, radically reducing the energy required for long-distance pumping.

Furthermore, I predict a global harmonization of effluent standards. As the link between micro-fiber inhalation or ingestion and pulmonary disease becomes clinically undeniable, absolute UF filtration will become the legally mandated minimum for any facility discharging into public waterways. The future of global water is a world where the polyethylene density trap is neutralized by the sheer mechanical precision of the steric blockade, ensuring that our synthetic legacy is captured and sequestered before it ever reaches the sea.

The Adaptive Sieve: Bio-Sedimentary Filtration and the Enzymatic Erasure of Polymers

The evolution of water remediation has transitioned from the static interception of particles to the engineering of living filtration matrices. As the global influx of low-density, non-polar polymers overwhelms legacy infrastructure, we are seeing the rise of a new paradigm: the utilization of the contaminant itself to facilitate its own removal. By architecting dynamic biological barriers and hybrid enzymatic reactors, we can achieve a level of aqueous purity that was previously considered thermodynamically impossible.

The DM Protocol: Engineering the 'Cake Layer' Defense

At the absolute frontier of low-energy filtration is dynamic membrane (DM) technology. Unlike traditional systems that rely on a synthetic, micro-porous screen, the DM protocol utilizes a secondary matrix composed of a highly controlled, physical accumulation of foulants—scientifically referred to as the cake layer. In this configuration, a cheap, high-porosity substrate (such as stainless steel mesh, woven dacron, or porous ceramic) acts merely as a skeleton. The true filtration work is performed by the dense, biological and particulate sediment that forms upon it.

This bio-sedimentary barrier creates an incredibly complex physical labyrinth that traps floating microplastics and nanoscale fibers. This represents a radical shift in engineering philosophy: rather than fighting membrane fouling, we manage it as a high-precision tool. However, this requires a delicate barotropic balance. If the layer becomes too compacted, it induces a state of hydraulic hindrance, where transmembrane pressure (TMP) skyrockets and the flux rate collapses.

Deep Analysis: The Physics of Porosity

The success of a DM depends on the precise manipulation of TMP and cross-flow velocity. By controlling these variables, engineers can maintain a steady-state cake thickness that maximizes the interception of neutrally buoyant polymers—like polyethylene (PE)—while ensuring a high volumetric flow. This technology is uniquely suited for developing regions and cost-conscious industrial hubs, as it bypasses the need for expensive chemical coagulants and the high-energy electrical inputs required by traditional high-pressure systems.

Comparative Filtration Performance

Feature	Sedimentation (Legacy)	Ultrafiltration (Standard)	Dynamic Membrane (Bio-Sieve)
PE Removal (< 0.5 mm)	15%	99%	92%
Energy Consumption	Low	High (Barotropic)	Low (Steady-State)
Pathogen Exclusion	Low	Absolute (Steric)	Variable (Biological)
Maintenance Cost	Low	High (Chemical Clean)	Low (Managed Fouling)

The Biological Siege of Additives

The true power of the MBR lies in its ability to strip away the chemical ghost of plastics. Billions of specialized bacteria and enzymes within the bioreactor target toxic leachates—specifically phthalate esters and bisphenols—breaking them down into benign metabolic byproducts. This biological purification of the plastic shards in the retentate stream prepares them for final molecular recycling or total mineralization.

Practical Advice for Industrial Remediation

To move from simple filtration to total erasure, facility operators should implement the following synergistic protocols:

- **The Enzyme-Membrane Nexus:** Do not rely solely on the physical sieve. To combat membrane blinding by hydrophobic plastics, the influent should be pre-treated with a cocktail of lipases and esterases. This initiates the surface softening of the polymer before it touches the membrane, ensuring that the cake layer remains both porous and enzymatically active.

- **HRT Optimization:** The hydraulic retention time (HRT) must be strictly managed to prevent toxic shock. If the initial chemical feed concentration of plasticizers is too high, it will collapse the bacterial colony. Gradual loading allows the microbes to adapt their genetic expression toward polymer catabolism.

- **Closed-Loop Reclamation:** Heavy industries, such as textiles, paper, and petrochemicals, should transition to internal recirculation. By installing MBRs directly at the point of generation, they can reclaim and reuse process water, turning a toxic liability into a circular, plastic-free asset.

Strategic Predictions: The Advent of 'Monomer Recovery' Refineries

I predict that within the next fifteen years, the holy grail of remediation will be realized through directed metabolic splicing. We will move beyond using wild bacteria toward the deployment of super-strains like a genetically optimized *Ideonella sakaiensis*. These organisms will be engineered to perform monomer recovery—the process of enzymatically deconstructing polyethylene terephthalate (PET) waste back into its foundational terephthalic acid and ethylene glycol within the treated effluent. Rather than filtering waste, the treatment plant will become a molecular refinery.

Furthermore, I predict the emergence of the nano-sieve interface. This will be a new class of bio-synthetic membranes that utilize aquaporins and protein channels to facilitate the zero-pressure passage of pure water while selectively capturing polymers at the atomic level. The future of the hydrosphere is one where the industrial errors of the 20th century are finally erased through a unified architecture of life and machine.

MBR vs. Conventional Treatment Comparison

Parameter	Conventional Activated Sludge	Membrane Bioreactor (MBR)
Effluent Quality	Variable (Turbidity common)	High (Pristine/Sterile)
Microplastic Removal	60% – 85%	> 99.9%
Footprint	Large (Requires Clarifiers)	Compact (Integrated)
Bacterial Concentration	Low to Moderate	High (High MLSS)
Pathogen Removal	Requires Chlorine/UV	Absolute (Physical Barrier)

The Molecular Gatekeeper: Nanocomposite Membranes and the Geopolitical Siege of Synthetic Flux

The remediation of the global hydrosphere has reached a state of material irony: we are currently forced to utilize advanced synthetic polymers to sequester the fragmented remains of 20th-century industrial polymers. However, the contemporary evolution of membrane science has moved beyond simple filtration. By integrating nanotechnology and systemic point-source interdiction, we are transitioning toward a state of aqueous sovereignty, where the flow of synthetic waste is halted at the molecular level before it can achieve environmental permanence.

1. The Nanocomposite Vanguard: Strengthening the Polymer Matrix

Polymeric membranes—often referred to as organic membranes—represent the frontline of pressure-driven separation. Utilizing sophisticated phase-inversion manufacturing, engineers can now dictate the absolute porosity of these filters down to the sub-nanometer scale. While materials like polytetrafluoroethylene (PTFE), polysulfone (PSf), and polyethersulfone (PES) provide high chemical resistance, their primary operational bottleneck remains biofouling—the accumulation of bacterial slime and hydrophobic plastic shards that blind the membrane.

Deep Analysis: The Reactive Interface

To overcome this, the next generation of filtration relies on nanocomposite embedding. By impregnating a polyvinylidene fluoride (PVDF) matrix with reactive metal oxides such as titanium dioxide (TiO2) or biocidal silver (Ag) nanoparticles, the membrane surface becomes a lethal environment for microbes. Furthermore, the inclusion of carbon nanotubes (CNTs) provides a structural reinforcement that increases the clean water flux while reducing the energy tax of high-pressure operations. We are no longer building passive sieves; we are engineering reactive gates that actively repel and dismantle the foulants that attempt to obstruct them.

2. The Geopolitical Siege: Point-Source Interdiction

The most strategically logical approach to the microplastic crisis is not the futile attempt to clean the ocean, but the absolute blockade of terrestrial waste. The marine environment functions as a kinetic grinder where crashing waves and high-intensity UV radiation shatter macro-debris into trillions of microscopic shards. This fragmentation is most aggressive on sandy beaches, which act as the primary generation zones for new microplastics.

The Arterial Defense: If we are to protect the marine biosphere, we must treat global river systems as the primary arteries of contamination. A coordinated, aggressive cleanup of heavily polluted river mouths and urban streams is the only logistical method for preventing buoyant waste from achieving oceanic distribution. This must be coupled with the mandatory legal requirement for all municipal wastewater treatment plants to install ultrafiltration (UF) or reverse osmosis (RO) polishing stages. By capturing the < 100 μm fraction at the source, we effectively sever the synthetic feedback loop.

3. Chemical Coalescence: The Silane Solution

For dynamic freshwater systems where mechanical filtration is logistically difficult, we are seeing the rise of chemical agglomeration. This technique utilizes specialized organosilane fixatives to manipulate the molecular physics of the water column. Microplastics are inherently difficult to capture due to their neutral buoyancy and small size. However, organosilanes utilize van der Waals molecular forces to create a targeted chemical affinity for synthetic polymers. When injected into a waste stream, these fixatives cause microplastics to violently bind together, forming massive, macroscopic clumps. This forced coalescence transforms an invisible, sub-micron threat into a buoyant, filterable mass that can be mechanically skimmed in a fraction of the time required for traditional filtration.

4. Practical Advice for Environmental Remediation Engineers

To move toward a state of total interdiction, I recommend the following tactical protocols:

- **The Brine Disposal Mandate:** The reject brine from RO desalination—which contains a hyper-concentrated load of micro- and nanoplastics—must never be discharged back into the sea. I recommend the integration of high-temperature plasma gasification. This process reduces the plastic brine to a stable, vitrified slag and a hydrogen-rich syngas, turning a toxic byproduct into a source of industrial energy.

- **Aggressive Beach Forensics:** Municipalities should implement high-frequency mechanical grooming of coastal sands. By removing macro-plastics before they undergo UV-induced photo-oxidation, we can reduce the localized generation of microplastics by an estimated 70%.

- **The Hydrophilic Shift:** Prioritize the use of cellulose acetate or graphene-oxide modified membranes for wastewater tertiary treatment. These materials possess a natural hydrophilicity that prevents the plastic-to-plastic bonding that characterizes the fouling of standard polyethylene filters.

5. Predictions: The Era of 'Smart Silanes' and Molecular Sovereignty

I predict that within the next decade, we will witness the emergence of smart silanes. These will be autonomous, biodegradable chemical agents designed to be released into coastal storm drains. They will act as molecular magnets, binding to plastics while ignoring organic matter, and will be programmed to float for exactly 24 hours for easy collection before they naturally degrade.

Furthermore, I predict the total obsolescence of open-sea cleaning. As the cost-efficiency of arterial interdiction (filtering rivers and wastewater) becomes undeniable, global funding will shift entirely toward land-based blockades. We will see the rise of zero-emission estuaries, where the mouth of every major river is equipped with an automated, solar-powered nanofiltration gate. The Plasticene will end not because we cleaned the oceans, but because we finally learned how to close the door.

The Membrane Paradox: Chemical Rehabilitation and the Circularity of Molecular Sieves

The global expansion of advanced water treatment has created a profound industrial contradiction. While high-performance membranes are the premier defense against microplastic infiltration, their own lifecycle constitutes a massive, secondary accumulation of synthetic waste. As we move toward a more health-conscious and resource-scarce world, the engineering community is forced to pivot from a linear consumption model to a circular siege of the materials themselves. The objective is no longer merely to filter water, but to implement the molecular refurbishment of the filters, ensuring that the act of purification does not leave an indestructible petrochemical footprint.

The Chemistry of Upcycling: From RO to UF Metamorphosis

The most technologically viable pathway for membrane circularity is the aggressive chemical metamorphosis of fouled units. In this paradigm, an exhausted reverse osmosis (RO) membrane is not viewed as waste, but as a pre-form for a secondary filtration matrix.

Membrane Lifecycle Transformation

Material State	Target Pore Size	Application	Functional Layer
Virgin RO	< 1 nm	Desalination / Nanoplastics	Polyamide (Active)
Fouled RO	Blocked	Disposal (Inefficient)	Scaling / Biofilm
Upcycled UF	10 – 100 nm	Tertiary Wastewater	Polysulfone (Support)

Practical Advice for Closed-Loop Membrane Management

To operationalize this molecular upcycling, water treatment facilities should adopt a "cascading utility" framework for their filtration assets:

- **Toxicity Scanning:** Before initiating chemical metamorphosis, spent RO membranes must undergo a forensic toxicity audit. Membranes exposed to persistent organic pollutants (POPs) or heavy metals may have these toxins embedded within the PSf matrix, making them unsuitable for reuse in potable water systems.

- **Controlled Oxidation:** The sodium hypochlorite exposure must be precisely calibrated. Excess concentration or prolonged contact time will not only strip the polyamide layer but will also begin to cleave the polysulfone structural bonds, leading to catastrophic mechanical failure of the membrane under pressure.

- **Secondary Market Creation:** Municipalities must create secondary markets for upcycled UF membranes. A filter that can no longer produce drinking water remains an exceptional tool for agricultural irrigation or industrial cooling water, replacing the need for virgin polymer production.

Predictions: The Rise of the 'Self-Regenerating' Sieve

I predict that within the next fifteen years, the "Static Sieve" will be replaced by the self-regenerating membrane interface. We will see the development of nanocomposite membranes where the active layer is designed to be sacrificial.

Upon fouling, a mild, non-toxic chemical flush will be used to trigger the in situ dissolution of the active layer, which will then be automatically redeposited by a companion microbial consortium living within the membrane module. This will create a state of infinite filtration, where the molecular gatekeeper is constantly renewing itself, using the captured carbon from the water column as the building blocks for its own reconstruction. The ultimate goal of the "Circular Siege" is a world where waste is not a landfill destination, but a thermodynamic pause on the road to synthetic resurrection.

The Kinetic Barrier: Infrastructural Obsolescence and the Synergy Pyramid of Global Recovery

The current global effort to mitigate microplastic pollution is fundamentally handicapped by infrastructural inertia. While modern society has mastered the removal of biodegradable organic matter, our legacy systems are structurally transparent to the non-polar, high-buoyancy synthetic particulates of the Plasticene. To move from passive contamination to active restoration, we must acknowledge that current municipal wastewater treatment plants (WWTPs) are functioning as concentrated point-source emitters. The resolution requires a total abandonment of antiquated settling methods in favor of a zero-pass membrane mandate.

1. The Infrastructural Paradox: Legacy Failure and the MBR Mandate

Standard wastewater treatment is designed for substances that either settle or decompose. Microplastics—and the exponentially more elusive nanoplastics—do neither. Consequently, billions of gallons of effluent are discharged daily with a synthetic load that bypasses primary and secondary clarifiers.

Deep Analysis: The Point-Source Paradox

The highest concentration of microplastics is not found in the open ocean, but at the discharge gates of our cities. Advanced tertiary treatment, specifically utilizing membrane bioreactor (MBR) technology, is the only current mechanical configuration capable of achieving approximately 100% physical exclusion of all suspended micro-shards. Beyond mere filtration, the MBR simplifies the entire treatment pipeline by consolidating biological aeration and physical separation into a single, high-efficiency unit.

2. The Logistics of Prevention vs. The Hallucination of Extraction

There is a dangerous, albeit popular, focus on open-ocean cleanup. From a logistical standpoint, attempting to extract microplastics in situ from the turbulent, vast marine environment is an exercise in futility. The energy expenditure and surface-area coverage required for such an operation are astronomically prohibitive.

Practical Engineering Advice: Terrestrial Interdiction

The most cost-effective and manageable strategy is terrestrial interdiction. By focusing on three critical blockade zones, we can sever the synthetic link:

- **Urban Arteries:** Mandating MBR for all municipal effluents to stop the flow at the source.

- **Coastal Forensic Cleaning:** Mechanical grooming of beaches where UV-induced fragmentation is most aggressive.

- **Riverine Gates:** Installing automated, high-surface-area filtration nets at major river mouths to trap buoyant waste before it achieves oceanic distribution.

3. The Synergy Pyramid: A Tri-Axis Framework for Global Recovery

The permanent resolution of the polymer crisis cannot be achieved through technology alone. It requires a synergy pyramid—a multi-dimensional framework that balances socioeconomic development with ecological sovereignty.

- **Axis 1: Advanced Membrane Science:** Transitioning to hyper-compact, enclosed WWTPs capable of processing all polymer types and sizes with a mandated 100% removal efficiency.

- **Axis 2: Legislative Purity:** Implementing molecular tariffs and highly punitive global policies that penalize the discharge of unfiltered effluents and the production of non-circular polymers.

- **Axis 3: Material Evolution:** The heavily subsidized mass production of biogenic polymers derived from natural sources that are 100% marine-degradable.

4. Predictions: The Era of the 'Enclosed Bio-Reactor'

I predict that within the next two decades, the open-air wastewater basin will become a relic of the past. Future WWTPs will be fully enclosed hyper-facilities. These will be compact, modular, and enzymatic-physical hybrids, capable of stripping water of all synthetic records and returning it to the water table in a state of pre-industrial purity.

Furthermore, I predict the rise of energy-independent desalination. As thermal and membrane technologies converge with renewable energy, we will see the birth of massive, solar-powered purity hubs that provide plastic-free hydration while operating entirely off the grid.

The Global Sovereignty of Purity: An Executive Strategy for the Total Neutralization of Synthetic Flux

The ecological survival of the 21st-century biosphere depends on a fundamental barotropic siege—the application of advanced membrane architectures to intercept the molecular-scale errors of our industrial past. We have reached a point where legacy wastewater infrastructure is no longer a defense, but a conduit for contamination.

1. The Hierarchy of Exclusion: Tactical Membrane Architectures

The remediation of synthetic waste requires a multi-modal engineering approach, where each membrane geometry plays a specific role in the deconstruction of the pollutant stream.

- **Reverse Osmosis (RO):** The barotropic siege. RO represents the brute-force physical limit of filtration. By applying a hydrostatic pressure of 10–100 bar, we override the natural entropic flow of water to achieve molecular-level separation. While highly efficient at capturing the nanoplastic fraction (< 100 nm), RO is susceptible to hydrophobic blinding.

- **Ultrafiltration (UF):** The steric blockade. UF provides a low-energy, high-volume solution for the physical exclusion of suspended solids and waterborne pathogens. By achieving a 95% reduction in biochemical oxygen demand (BOD), UF allows for the industrial reclamation of process water.

- **Dynamic Membranes (DM):** The bio-sedimentary filter. DM technology utilizes the cake layer—a controlled accumulation of biological and particulate matter—as its primary sieve. This adaptive sieve is uniquely suited for the decentralized treatment of neutrally buoyant plastics like polyethylene (PE).

- **Membrane Bioreactors (MBR):** The apex of remediation. The MBR is the definitive hybrid of life and machine. By coupling high-density membrane separation with the enzymatic power of specialized bacterial consortia, the MBR achieves a staggering 99.9% removal efficiency.

2. The Molecular Refurbishment: Circularity and Upcycling

The membrane paradox—the use of synthetic polymers to filter synthetic polymers—must be resolved through a state of molecular circularity. We must move away from the linear consumption of filter media toward a strategy of chemical upcycling.

3. Practical Advice for Global Stakeholders

To achieve aqueous sovereignty, I recommend the following mandates for recovery:

- **Arterial Interdiction:** Focus cleanup efforts on the primary arteries—major river mouths and urban storm drains. It is logistically impossible to clean the open ocean; we must instead close the gates at the terrestrial source.

- **Enzymatic Priming:** Industrial-scale bioreactors should implement enzymatic pre-shocks using proteinase K or lipase cocktails to soften plastic waste before it touches the membrane.

- **Brine Gasification:** Reject-brine from desalination plants, which contains concentrated nanoplastics, must be processed via high-temperature plasma gasification rather than deep-sea discharge.

4. Predictions: The Era of 'Smart Porosity' and Bio-Sovereignty

I predict that within the next two decades, we will witness the death of the static membrane. Utilizing nanotechnology and smart polymers, we will see membranes with dynamic pore architectures that adjust their geometry in real-time to match the particulate signature of the water.

Comparison of Advanced Filtration Technologies

Technology	Pressure (bar)	Pore Size	Primary Mechanism	Microplastic Removal
MBR	0.1 – 0.5	0.04 – 0.1 μm	Bio-Physical Hybrid	> 99.9%
UF	1 – 10	1 – 100 nm	Steric Exclusion	> 99%
RO	10 – 100	< 1 nm	Solution-Diffusion	Absolute
DM	< 0.1	Variable	Bio-Sedimentary	90% – 95%

The Synthetic Metabolism: Reprogramming the Microbial Secretome for Global Polymer Erasure

The contemporary biosphere is currently saturated with an estimated 100 million tons of raw plastic waste annually—a volume that natural evolutionary processes are fundamentally unequipped to manage. With polyethylene (PE) and polypropylene (PP) constituting approximately 92% of this synthetic influx, the planet's primary ecological sinks have been converted into permanent petrochemical repositories. Traditional waste management—ranging from the energetic inefficiency of mechanical downcycling to the toxic exhaust of mass incineration—has reached a thermodynamic dead end. To resolve this, we must pivot from the observation of natural decay to the revolutionary bioengineering of a synthetic metabolism, utilizing the tools of systems biology to turn industrial waste into a biological feedstock.

The Biological Bottleneck: Why Evolution is Failing

Natural biodegradation is a process of evolutionary lag. While certain insects, bacteria, and fungi have developed a latent capacity to graze on polymers, their metabolic rates are orders of magnitude too slow to counteract the anthropogenic deluge. This is due to the crystalline barrier of PE and PP, where the tight molecular packing of the carbon-carbon backbone resists enzymatic penetration.

Deep Analysis: The Genomic Short-Circuit

The failure of natural microbes to degrade plastic at scale is not a lack of intent, but a lack of enzymatic torque. Most wild-type enzymes lack the specific binding affinity required to anchor onto a hydrophobic, chemically inert surface. To overcome this, we must short-circuit the slow march of natural selection. By utilizing computational genomics, we can identify the specific gene clusters responsible for polymer catabolism in extremophiles and reprogram them into high-growth industrial hosts.

The Rise of the Bio-Refinery: Upcycling the 'Plasticene'

The ultimate goal of bioengineering is not merely the destruction of waste, but its molecular upcycling. We are moving toward a closed-loop industrial model where highly toxic microplastics are no longer viewed as pollutants, but as a high-density carbon source (C-source) for the production of advanced biopolymers. In this scenario, a reprogrammed microbe ingests a PET or PE fragment and, through a series of engineered intracellular pathways, converts that carbon into polyhydroxyalkanoates (PHA) or other medical-grade bioplastics.

Practical Advice for Synthetic Biologists and Engineers

To achieve hyper-efficient degradation, the focus must shift from single-enzyme studies to pathway optimization:

- **The Multi-Enzyme Chassis:** Do not rely on a single PETase or lipase. To dismantle HDPE, the microbial host must be engineered with a synergistic secretome that includes alkane monooxygenases for initial oxidation and hydrolases for subsequent chain cleavage.

- **Hydrophobic Anchoring:** Genetically modify the cell surface of the host microbe to express adhesion proteins. This ensures the microbe remains physically locked onto the plastic shard, creating a localized high-concentration zone of enzymes.

- **Metabolic Redirecting:** Utilize CRISPR-i (interference) to shut down non-essential metabolic pathways in the host. This forces the microbe to dedicate 100% of its cellular energy to polymer deconstruction and the synthesis of high-value metabolites.

Architecting the Synthetic Scavenger: Genomic Editing and the Enzymatic Siege

The biological deconstruction of synthetic polymers requires a departure from natural evolutionary timelines. While the planet has managed biopolymers for eons through well-established carbon cycles, the introduction of high-density polyethylene (PE) and polypropylene (PP) created a kinetic lock that wild-type organisms cannot easily pick. The contemporary bioengineering frontier is defined by the effort to find, sequence, and hyper-accelerate the latent enzymatic potential found in extremophile genera.

1. The Wild-Type Surveillance: Identifying the Bio-Catalytic Baseline

Nature has already begun its slow, adaptive response to the Plasticene. A specialized cadre of microbial genera—specifically *Bacillus*, *Pseudomonas*, *Rhodococcus*, and *Stenotrophomonas*—has been identified as the evolutionary baseline for polymer catabolism. These organisms have learned to treat the synthetic carbon-carbon backbone as a primary energy vector (C-source), though their natural rates of degradation remain commercially non-viable.

Deep Analysis: The Mycelial Drill and the Bacterial Scythe

The deconstruction of highly crystalline polymers is a dual-force operation. Fungal strains like *Alternaria alternata* and *Penicillium* species act as biological drills, utilizing laccases and peroxidases to physically penetrate the polymer matrix. Simultaneously, bacterial isolates like *Stenotrophomonas pavanii* provide the scythe, utilizing targeted extracellular enzymes to harvest the fragments

2. The PETase-MHETase Cascade: A Molecular Tag-Team

The most significant breakthrough in directed remediation is the discovery and subsequent reprogramming of the *Ideonella sakaiensis* enzymatic pathway. This organism provides a blueprint for what is known as the molecular tag-team:

- **Phase I: PETase Infiltration.** The extracellular enzyme PETase targets the dense polyethylene terephthalate (PET) chain, performing a violent hydrolysis that yields a reaction intermediate: mono(2-hydroxyethyl) terephthalic acid (MHET).

- **Phase II: MHETase Cleavage.** A second specialized enzyme, MHETase, instantly targets the intermediate, reducing it to its foundational, harmless monomers: terephthalate and ethylene glycol.

3. Polyurethane Deconstruction: The Aromatic Breach

Polyurethanes (PU) represent a particularly complex challenge due to their diverse chemical linkages. Specialized strains of *Pseudomonas* have shown the ability to utilize PU-diol and toxic intermediates like 2,4-diaminotoluene as metabolic fuel. The resulting metabolic exhaust—volatile amines, organic acids, and alcohols—represents the final stage of the aromatic breach, where the most durable synthetic foams are returned to basic organic markers.

4. Practical Advice for Bioprocess Engineering

- **The 'Synthetic Secretome' Cocktail:** Do not rely on a single strain. For LDPE and PU remediation, deploy a synergistic consortium. This creates a multi-front assault that can increase mass-reduction rates by 13–20% over single-strain applications.

- **Thermophilic Priming:** In high-heat industrial reactors, utilize hyper-robust strains like *Brevundimonas* and *Sphingobacterium*. These thermophilic scavengers remain enzymatically active at temperatures that would denature standard proteins.

- **In Silico Enzyme Folding:** Prior to genetic splicing, use AI protein-folding models to hyper-stabilize the PETase and MHETase enzymes. Increasing the thermal stability of these proteins allows them to function in the harsh environments of municipal waste slurries.

Comparison of Natural vs. Engineered Remediation

Parameter	Natural Evolution (Wild-Type)	Synthetic Metabolism (Engineered)
Degradation Rate	Years to Decades	Hours to Weeks
Crystallinity Resistance	High (Ineffective)	Low (Engineered Penetration)
End Product	Micro-fragments	Biogenic Feedstock (PHA)
Temperature Tolerance	Narrow (Ambient)	Wide (Thermophilic Optimization)

The Kinetic Fortress: Thermodynamic Engineering and the AI-Driven Enzymatic Leap

The transition from laboratory observation to industrial-scale plastic erasure is currently obstructed by a kinetic fortress. Wild-type enzymes, while biologically fascinating, are structurally inadequate for the high-friction, high-heat environments of industrial bioreactors. They suffer from low catalytic torque, extreme substrate rigidity, and a catastrophic tendency to denature when removed from their native cellular niches. To breach this fortress, we must move beyond natural selection and utilize computational protein engineering to architect enzymes that are not only faster but thermodynamically invincible.

1. Thermodynamic Hardening: The Disulfide Siege

The primary bottleneck in enzymatic remediation is thermal instability. Industrial degradation generates significant localized friction and metabolic heat; an enzyme that denatures at 40°C is a liability. To harden these proteins, bioengineers are now utilizing rational design to replace weak hydrogen bonds with robust disulfide bridges (S-S).

2. The Transgenic Host Revolution: Living Reactors

The second bottleneck is the secretory limit. For remediation to be scalable, we must move the enzymatic factory into robust, high-growth hosts.

- **Yeast Scaffolding:** The industrial yeast *Yarrowia lipolytica* has been successfully reprogrammed to serve as an extracellular pump, continuously secreting cutinase derived from *Fusarium solani*. This allows for rapid, mass fermentation where the living liquor maintains stable PET degradation at 28°C without the need for expensive purified enzyme cycles.

- **The Algal Vanguard:** In a profound leap for marine remediation, PETase sequences have been integrated into the marine microalga *Phaeodactylum tricornutum*. This creates a self-remediating algal bloom, where the microalgae not only photosynthesize but actively dissolve high-crystallinity PET shards in the surrounding water column.

Enzyme and Polymer Performance Metrics

Metric	Wild-Type Enzyme	AI-Optimized Enzyme	PET (Petrochemical)	PHA (Biogenic)
Melting Temp (Tm)	40°C – 45°C	55°C – 60°C	N/A	N/A
Catalytic Rate	Baseline (1x)	40x	N/A	N/A
Marine Half-Life	N/A	N/A	400+ Years	6 – 12 Months
Degradation Mode	Surface Only	Deep Penetration	Inert	Bio-Active

3. AI-Driven Mutation:

The most disruptive force in bioengineering is the fusion of AI-driven mutation design and supercomputing. By utilizing systematic genetic clustering and sequence alignment simulations, researchers have bypassed millions of years of trial-and-error evolution. Computational models can now predict the absolute preferred amino acid sequences required to maximize degradation throughput. This has resulted in enzymatic variants demonstrating a exponential increases increase in catalytic activity compared to wild-type baselines.

Practical Advice for Bioprocess Engineers

- **The Disulfide Mandate:** Every industrial enzyme candidate must be audited for thermal flex. If the Tm is below 55°C, the protein must be reinforced with disulfide bridges to prevent catastrophic denaturation during high-shear cycles.

- **Electrostatic Attraction Engineering:** Utilize computational mapping to modify the enzyme's surface charge. By making the enzyme surface more electrostatically compatible with the target polymer, you can increase the binding velocity.

- **Host-Specific Secretion Signals:** When using *Yarrowia* or *E. coli*, optimize the signal peptides to ensure the maximum percentage of synthesized enzymes are successfully secreted into the medium.

The Autonomous Evolution: AI-Governed Bioreactors and the Multi-Omic Siege

The integration of advanced microbial engineering into plastic waste management has revealed a profound translational bottleneck. While transgenic strains demonstrate phenomenal performance in sterile, 30°C laboratory environments, they frequently suffer catastrophic failure when deployed in high-shear, open-environment industrial reactors. The pressures of viral bacteriophage attacks, fluctuating thermal parameters, and intense ecological competition create a biological noise that stifles engineered pathways. To overcome this, we must pivot from static genetic edits to the deployment of AI-governed autonomous evolution.

1. The Multi-Omic Arsenal: Systems-Level Interrogation

To bridge the lab-to-field chasm, we must move beyond single-gene monitoring. The next generation of bioengineering requires a multi-omic siege—the simultaneous, real-time integration of high-throughput genomics, transcriptomics, metabolomics, and proteomics.

Deep Analysis: The Metabolic Flux Audit

By utilizing systems biology and bioinformatic characterization, engineers can perform a metabolic flux audit on recombinant strains. This allows us to identify exactly where the carbon flux is being diverted away from polymer deconstruction and toward survival-mode cellular maintenance. Through CRISPR-Cas9 full-genome editing, we can then lock the desired traits, effectively forcing the microbe to prioritize enzymatic secretion even under the stress of industrial-scale competition.

2. The AI Blueprint: Unraveling the Nanoscale Interface

The physical interaction between a microbial cell and a hydrophobic plastic surface takes place at the nanoscale. Unraveling these complex dynamics requires the aggressive integration of machine learning (ML) and functional metagenomics. Digital twin models can now provide a mathematically optimized blueprint for protein engineering. By simulating millions of potential amino acid substitutions, AI can predict which mutations will enhance the binding affinity of an enzyme to a weathered, UV-degraded polymer surface.

3. Practical Advice for Bioinformaticians and Systems Engineers

- **The Weathering Mandate:** Neural networks must be trained on weathered datasets. If you train a model strictly on the structure of virgin polymers, the resulting enzymes will fail in the wild.

- **Phage-Resistance Engineering:** When designing industrial hosts, implement CRISPR-Cas immunity systems specifically targeted against common wastewater bacteriophages to prevent viral collapse.

- **Closed-Loop Cradle-to-Cradle Design:** The remediation pipeline must begin in the design phase. Manufacturers should prioritize the synthesis of polymers that are pre-programmed with low-energy biological processing pathways in mind.

4. Strategic Predictions: The Advent of 'Directed Evolution Reactors'

I predict that within the next decade, the static strain model of bio-remediation will be obsolete. We will witness the rise of AI-governed directed evolution reactors. These automated systems will not rely on a single engineered bacterium. Instead, they will be evolutionary pressure cookers that continuously mutate and sequence microbial populations in real-time. The AI will monitor the degradation efficiency and select the most successful variants for immediate clonal expansion.

The Bio-Architectural Reconquest: Synthetic Biology as the Terminal Solution to Polymer Recalcitrance

The global saturation of polyethylene (PE) and polypropylene (PP) represents a challenge to the fundamental kinetic limits of the carbon cycle. These polymers, designed for absolute structural inertia, have effectively locked billions of tons of carbon into an indestructible state. However, we are currently witnessing the dawn of the bio-architectural reconquest. By identifying the evolutionary elite and subjecting them to aggressive genomic reprogramming, we are moving toward the total biological erasure of synthetic waste.

1. The Elite Vanguard: Taxonomical Assets for Deconstruction

Nature's response to the polymer influx has produced a specialized microbial vanguard. While wild-type organisms lack the speed required for industrial-scale remediation, their latent genetic blueprints provide the raw materials for bioengineering.

- **Bacterial Scavengers:** Genera such as *Pseudomonas*, *Bacillus*, and *Staphylococcus* have demonstrated a remarkable ability to adapt their metabolic flux toward the carbon-carbon backbone of PE and PP.

- **Fungal Penetrators:** *Aspergillus* and *Penicillium* utilize high-torque extracellular enzymes, including laccases and peroxidases, to physically and chemically breach the crystalline fortress of high-density plastics.

2. The Three Pillars of Metabolic Programming

To accelerate these natural processes into an industrial-scale biological siege, we utilize three distinct engineering levers:

- **Rational Protein Engineering:** Utilizing AI-driven computational models to architect super-enzymes. By inserting synthetic disulfide bridges (S-S), we increase the thermal stability (Tm) of these proteins.

- **Genomic Gene Shuffling:** A technique of molecular recombination where beneficial mutations from multiple strains are shuffled into a single, hyper-efficient chassis.

- **Aggressive Directed Evolution:** Placing microbial colonies in evolutionary pressure cookers where the only available energy source is the target polymer, forcing the evolution of hyper-specific enzymatic keys.

3. Practical Advice for the Next Generation of Bio-Architects

- **The Cradle-to-Cradle Mandate:** Future polymer manufacturing must be inextricably linked to bioengineering. Plastics should be engineered with enzymatic trigger sites—specific molecular weak points pre-matched to our engineered microbes.

- **Consortia Design:** Do not engineer isolated strains. The most resilient remediation occurs in synthetic biofilms—carefully architected multi-species consortia where the metabolic exhaust of one species serves as the fuel for the next.

- **Metabolic Redirecting:** Utilize CRISPR-Cas9 to disable non-essential pathways in your production host, forcing the cell to dedicate 100% of its intracellular ATP to enzyme synthesis.

Comparative Analysis of Bio-Remediation Strategies

Strategy	Primary Mechanism	Implementation Scale	Potential Efficacy
Wild-Type Grazing	Natural Selection	Global (Passive)	Low (Evolutionary Lag)

Strategy	Primary Mechanism	Implementation Scale	Potential Efficacy
Static Genetic Edits	CRISPR-Cas9	Laboratory (Pilot)	Moderate (Translational Gap)
Autonomous Evolution	AI-Governed Reactors	Industrial (Scale)	High (Real-Time Adaptation)
Synthetic Consortia	Metabolic Synergism	Ecosystem (Targeted)	Extreme (Total Mineralization)

Forensic Oceanography: Establishing the Global Protocol for Synthetic Detection

The silent invasion of synthetic particulates has necessitated a revolutionary shift in environmental forensic science. As heavy industry, global logistics, and large-scale construction continue to hemorrhage polymers into the biosphere, the scientific community is racing to move beyond mere observation toward a rigorous, standardized system of quantification. To sever the link between industrial production and ecological decay, we must first master the art of detection. Without a globally harmonized set of analytical protocols, the data generated by disparate laboratories remains a collection of isolated anecdotes rather than a cohesive map of a planetary crisis. The objective is to move toward a "Universal Language of Measurement" that allows policymakers to transition from vague concern to legally enforceable, data-driven intervention.

The Gradient of Capture: From Macroscopic to Microscopic

The hunt for synthetic debris begins with a strategic choice of resolution. Depending on the environment—whether it be the sun-drenched surface layer or the lightless abyssal floor—researchers deploy a hierarchy of extraction strategies.

The first level is Targeted Visual Reconnaissance. This involves the manual vetting and physical isolation of visible fragments, typically those occupying the one-to-five-millimeter spectrum. While this approach relies on human visual acuity and is most effective on coastal shorelines, it is inherently limited. The human eye is easily deceived; fragments often camouflage themselves within natural organic debris or become encrusted in mineral deposits. Consequently, this method is increasingly viewed as a preliminary screening tool rather than a definitive census.

For environments where the synthetic load is hidden or overwhelmingly microscopic, the protocol shifts to Total Mass Acquisition. Here, researchers collect the entire, unadulterated volume of a substrate—be it a liter of seawater or a kilogram of sediment—without initial filtration. This "bulk" approach ensures that even the most elusive nanoplastics are captured, though it demands significant laboratory processing to separate the synthetic needle from the organic haystack.

Finally, In-Situ Concentration represents the technological edge of field research. Rather than transporting massive volumes of water, researchers utilize high-precision, fine-mesh filtration systems to concentrate particles directly at the source. This volume-reduction process allows for the sampling of thousands of liters of seawater in a single session, providing a statistically robust snapshot of the ambient polymer density.

The Technological Arsenal: Skimming the Surface and Piercing the Deep

Capturing a representative sample of the world's oceans requires a diverse array of specialized hydrodynamic tools. For the surface microlayer, researchers deploy "Manta Trawls" and "Neuston Catamarans"—aerodynamic gliders equipped with wide intake mouths and fine-mesh netting. These devices must be meticulously stabilized to prevent "wave hopping," a phenomenon that can disrupt the filtration flow and skew quantitative data. A critical component of this assembly is the mechanical flowmeter, a device that records the exact volume of water processed. This metric is the bedrock of normalization, allowing scientists to calculate precise concentrations of synthetic matter per cubic meter.

As the search moves vertically downward, the equipment becomes more robust. Bottom trawls and weighted grabbers are utilized to harvest the heavy benthic sediments that act as the planet's final plastic sink. In the extreme depths, where human presence is impossible, the protocol shifts to the use of Remotely Operated Vehicles (ROVs). These submersibles, capable of reaching depths of 5,000 meters, are equipped with surgical-grade mechanical manipulators and multi-coring cylinders to extract pristine samples from the abyssal strata.

Substrate Analysis: Life as a Biological Proxy

Perhaps the most complex dimension of detection is the extraction of polymers from biological tissues. Because marine life—from primary producers like zooplankton to apex predators—actively ingests these materials, the organisms themselves have become living indicators of environmental health.

The sampling protocol for biology is determined by the organism's ecological role. For smaller filter-feeders like mussels or barnacles, the entire specimen is typically harvested and transported. In contrast, for larger species such as cetaceans or predatory fish, researchers perform targeted surgical removals of the gastrointestinal tract. This process requires extreme sterility; the use of non-plastic tools and 100% natural fiber clothing is non-negotiable to prevent "procedural cross-contamination," where modern synthetic fibers from the researcher's own environment pollute the sample.

On the shoreline, the process is more manual but no less rigorous. Using standardized "quadrat" frames and cylindrical corers, researchers demarcate specific scientific zones to extract the top five centimeters of sediment—the layer most exposed to tidal influx. This meticulous attention to depth and area allows for the normalization of data across different continents, turning a simple beach survey into a piece of a global jigsaw puzzle.

Predictions: The Shift Toward Automated Forensics

As we look toward the next decade of detection, the reliance on manual sorting and physical netting will likely diminish. We are approaching a "Forensic Threshold" where the sheer volume of microplastics in the environment demands automated, real-world monitoring. I predict that the next generation of detection will utilize In-Situ Spectroscopic Sensors—autonomous devices capable of identifying the chemical "fingerprint" of a polymer the moment it passes through a sensor, transmitting data via satellite in real-time. This would eliminate the delay between field collection and laboratory analysis, allowing for an "Early Warning System" for synthetic surges.

The ultimate goal of these methodological frameworks is the transition from observation to elimination. By establishing a rigid, transparent, and universally accepted protocol for detection, we are essentially building the evidence base for a global treaty. The future of the "plastisphere" will be determined not just by what we produce, but by our ability to see and measure the invisible cost of our current civilization.

The Molecular Pivot: Biopolymer Synthesis and the Engineering of Ecological Reciprocity

The contemporary reliance on petrochemical polymers has created a structural stasis within the global carbon cycle. While human society is intrinsically tethered to the convenience of plastics, the absolute engineered resistance of the carbon-carbon (C-C) backbone has turned industrial utility into geological permanence. To resolve this, we must execute a molecular pivot—transitioning from materials designed for indifference to those designed for reciprocity. The biopolymer revolution is not merely a search for green alternatives; it is the strategic engineering of macromolecules that are chemically recognizable by the microbial world.

1. The Matrix of Recognisability: Why Biopolymers Succeed

The fundamental failure of conventional plastics—such as PE, PP, and PVC—is their chemical anonymity. The microbial secretome has no evolutionary experience with long-chain, highly stable synthetic hydrocarbons. In contrast, biopolymers are composed of monomeric units covalently bonded through linkages, such as esters, amides, and glycosidic bonds, that are ubiquitous in nature.

Deep Analysis: Enzymatic Recognition and Stoichiometry

True biopolymers, whether polyhydroxyalkanoates (PHAs) or polylactic acid (PLA), possess enzymatic trigger sites. Because their chemical structures mimic natural proteins and polysaccharides, they are susceptible to rapid, total mineralization. This is a stoichiometric victory: the polymer is treated as a high-density C-source for cellular metabolism, yielding CO_2, H_2O, and microbial biomass rather than fragmented microplastics.

2. The Ecological Imperative: Neutralizing the 'Microplastic Storm'

The adoption of biopolymers is a direct response to the microplastic storm—the aerosolized and aqueous circulation of synthetic particles that now infiltrate every biological tissue on the planet. Conventional plastics leach endocrine-disrupting chemicals and act as concentrated vectors for heavy metals. By utilizing drop-in biopolymer alternatives, we can maintain industrial output while ensuring that environmental release results in biological integration rather than systemic toxicity.

3. Practical Advice for Industrial Manufacturing and Transition

- **Processing Temperature Management:** Most biopolymers have narrower thermal processing windows than PE. To prevent thermal scission during injection molding, precisely calibrate extruder heat zones and utilize specialized bio-additives that prevent premature chain degradation.

- **The Moisture Sensitivity Barrier:** Biopolymers are, by definition, hydrophilic or susceptible to hydrolysis. Ensure that raw bio-pellets are stored in desiccant-controlled environments to prevent in-barrel degradation, which can lead to catastrophic losses in mechanical properties.

- **Additive Optimization:** When transitioning, avoid using standard petrochemical plasticizers. Utilize bio-derived citrate or sebacate esters to maintain flexibility while preserving the compostable status of the material.

The Proteomic Blueprint: Carbon Flux and the Engineering of Proteinaceous Matrices

The transition from the Plasticene to a circular bio-economy requires an intimate understanding of the metabolic origins of our materials. Unlike the toxic, refined monoliths of the petroleum industry, natural macromolecules are the products of billions of years of iterative evolutionary design. By mining the biological wealth of the pedosphere and the hydrosphere—ranging from plant-based structural proteins to the fibroin of silk—we are entering an era of adaptive material science.

1. Atmospheric Flux: The Carbon Accounting of Bio-Architectures

The fundamental difference between a bio-based polymer and a petrochemical plastic is a matter of temporal sequestration. Petrochemicals vent prehistoric carbon—carbon that has been sequestered for millions of years—directly into the modern atmosphere, permanently altering the CO_2 equilibrium. A bio-based polymer operates on a zero-net principle. The carbon used to construct a starch-based film or a proteinaceous matrix was drawn from the atmosphere via photosynthesis within the last 12–24 months.

2. The Proteomic Engine: Structural Complexity and Amide Linkages

Proteins represent the absolute apex of natural biopolymer engineering. Built from 20 distinct amino acids linked via covalent peptide (amide) bonds, proteins offer a degree of 3D-folding complexity that synthetic polymers cannot match.

- **The Amide Advantage:** Synthetic polymers like PE or PP rely on inert carbon-carbon (C-C) backbones. Proteins utilize amide linkages that provide sites for hydrogen bonding, allowing the material to switch between rigid and elastic states based on its folding geometry.

- **Structural Diversification:** By leveraging amino acid sequence specificity, engineers can develop polymers that are hydrophobic, such as zein from corn, or hyper-elastic, such as elastin or gluten.

3. Collagen and Gelatin: The Triple-Helix Matrix

Collagen is the foundational structural protein of the vertebrate world. Its mechanical integrity is derived from a unique, incredibly dense triple-helix structure.

- **Advanced Therapeutics:** In medical sciences, purified collagen is the premier choice for tissue engineering scaffolds and bio-absorbable sutures due to its high tensile strength and biocompatibility.

- **Gelatin Metamorphosis:** Gelatin is a partially hydrolyzed form of collagen. Both Type A (acidic pretreatment) and Type B (alkali pretreatment) are dominated by glycine, proline, and the critical 4-hydroxyproline residues, which grant the material its unique thermoreversible gelling properties.

4. Silk Fibroin: Beta-Sheet Innovation

Silk, primarily derived from the *Bombyx mori* silkworm, is a protein-rich biopolymer dominated by the protein fibroin. Its strength is a result of highly crystalline beta-sheet structures.

- **The Regenerative Miracle:** Purified silk fibroin possesses innate anticoagulation properties, preventing blood platelet adhesion. This makes it a miracle material for coating medical implants.

- **Bio-electronics:** Due to its dielectric properties and mechanical resilience, silk is increasingly being explored as a substrate for implantable, biodegradable bio-electronics.

Metric	Petrochemical (PE/PP)	Biogenic (PHA/PLA)	Proteomic (Silk/Collagen)
Carbon Origin	Prehistoric (Fossil)	Modern (Atmospheric)	Modern (Metabolic)
Primary Bond	Carbon-Carbon (C-C)	Ester	Amide (Peptide)
Reciprocity	Indifferent (Static)	Reciprocal (Degradable)	Adaptive (Bio-Active)
Industrial Use	Bulk Packaging	Compostable Goods	Advanced Med-Tech

The Carbohydrate Scaffold: Engineering Polysaccharide Architectures for Post-Petrochemical Industrialism

The transition from the Plasticene to a circular bio-economy requires an intimate understanding of the structural diversity of polysaccharides. These carbohydrate-derived plastics leverage the abundance of plant starch, structural cellulose, and marine-derived chitin to create high-volume, low-cost alternatives to PE (polyethylene) and PP (polypropylene). Unlike the toxic monoliths of the petroleum industry, these macromolecules are the products of iterative evolutionary design, offering a zero-net carbon profile and total biological integration.

1. Starch-Based Thermoplastics (TPS): The Plasticizer Siege

Native agricultural starches are among the most abundant and renewable raw materials on Earth. However, in their natural state, they are industrially non-viable due to high brittleness and poor thermal stability.

Deep Analysis: Hydrogen Bond Disruption

The conversion of native starch into thermoplastic starch (TPS) is a thermodynamic victory. By applying high heat and mechanical shear in the presence of low-molecular-mass plasticizers—such as glycerol or sorbitol—engineers can violently disrupt the tight intermolecular hydrogen bonds within the starch granule. This process lowers the glass transition temperature (Tg), transforming a crystalline powder into a moldable, flexible matrix.

Practical Advice for Material Scientists: The primary hurdle in TPS manufacturing is retrogradation—the re-crystallization of starch chains over time. To mitigate this, utilize a synergistic ratio of glycerol and sorbitol. This blend acts as a molecular wedge, locking the starch in an amorphous state and extending the shelf-life of compostable packaging.

2. Cellulosic Architectures: The Nanoscale Crystalline Breakthrough

Cellulose is the primary structural polysaccharide providing rigidity to all global plant life. As the most abundant organic polymer on the planet, its chemical derivatives are already essential to the food packaging and pharmaceutical industries.

Deep Analysis: The Nano-Fibrillation Protocol

To successfully replace transparent PE films, we must move beyond bulk cellulose toward nanocrystalline cellulose (CNCs). By mechanically grinding cellulose down to its nanoscale crystalline fibrils, we create a material with phenomenal oxygen barrier properties and absolute optical transparency. At the nanoscale, these fibrils provide a surface area-to-volume ratio that allows for superior mechanical reinforcement when blended with other biopolymers like PLA.

3. Chitin and Chitosan: Harnessing Invertebrate Armor

Second only to cellulose in sheer global biomass, chitin is a nitrogen-bearing structural polysaccharide that provides the armor-like rigidity of crustacean exoskeletons. Its derivative, chitosan, obtained via the aggressive deacetylation of chitin, is a miracle material for the biomedical sector.

Deep Analysis: Cationic Bonding Mechanics

Chitosan is unique among polysaccharides for its powerful cationic (positive) charge. In a physiological environment, this charge allows chitosan to bond aggressively with the negatively charged membranes of red blood cells and bacteria. This makes it an unparalleled tool for:

- **Hemostatic Response:** Instantly halting arterial bleeding by inducing rapid platelet aggregation.

- **Antimicrobial Action:** Disrupting bacterial cell walls on contact, preventing secondary infections in trauma care.

The Autotrophic Sieve and the Forensic Logic of Decay: PHAs and the Compostability Paradox

The transition to a post-petrochemical economy relies on two critical pillars: the biogenesis of high-performance polymers and the forensic verification of their end-of-life cycles. Polyhydroxyalkanoates (PHAs) represent the absolute frontier of bacterial polyester engineering, offering a drop-in mechanical replacement for PP. However, the success of these materials is currently hindered by a linguistic crisis—the deceptive marketing of biodegradability.

1. The PHA Platform: From Waste-Streams to Autotrophic Factories

PHAs are intracellular carbon-storage granules synthesized by specialized bacteria. Unlike starch-based plastics, PHAs possess high moisture resistance and a tensile strength that rivals conventional synthetics.

Deep Analysis: The Shift to Photoautotrophy

The primary bottleneck for PHA commercialization is the high cost of organic carbon feedstocks like glucose. To achieve industrial viability, we must bypass heterotrophic synthesis in favor of photoautotrophic microalgae.

- **Metabolic Efficiency:** By utilizing specialized microalgae that metabolize ambient solar light and atmospheric CO_2, we eliminate the requirement for expensive sugar feedstocks.

- **Carbon Sequestration:** This turns the plastic factory into a carbon sink. The polymer is literally mined from the atmosphere, ensuring that its eventual degradation adds zero net greenhouse gases to the cycle.

2. The Forensic Logic of Decay: Biodegradable vs. Compostable

There is an urgent need to dismantle the greenwashing of polymer degradation. To the engineer, it is a matter of kinetics and thermal thresholds. A strictly compostable item is always fundamentally biodegradable, but a broadly biodegradable material is not always compostable.

- **Biodegradability:** A generic term indicating that a material will eventually degrade via microbial action. Without a peer-reviewed timeframe (typically < 365 days), this claim is increasingly classified as deceptive.

- **Compostability (ASTM D6400):** This is a hyper-specific, legally enforced subset of degradation. It requires the material to decompose under managed industrial conditions: sustained moisture and thermophilic temperatures frequently exceeding 60°C.

3. Practical Advice for Industrial Compliance and Design

- **The One-Year Rule:** Do not market a product as degradable if it enters the solid waste system and does not fully decompose within 365 days. Failing this standard invites severe punitive legal action and greenwashing litigation.

- **Thermal Trigger Identification:** When designing PLA or PHA blends, clearly label the required degradation environment. If your material requires industrial heat (> 55°C), it must not be marketed as home compostable.

- **Metabolic Feedstock Optimization:** For facility engineers running PHA reactors, prioritize *Halomonas* strains. These can be grown in hyper-saline wastewater, reducing the need for sterile conditions.

Comparison of Industrial and Home Compostable Standards

Metric	Industrial Compostable (ASTM D6400)	Home Compostable	Marine Degradable
Temperature Requirement	Thermophilic (55°C – 60°C)	Ambient (20°C – 30°C)	Cold (0°C – 15°C)
Timeframe	180 Days	365 Days	Variable / Unstable
Primary Mechanism	Hydrolysis / Microbial	Microbial / Fungal	Microbial / Hydrolysis
Typical Material	PLA / High-MW Blends	PHA / Starch / Cellulose	PHB / Certain Esters

The Microbial Foundry: Mycelial Composites and the Synthesis of High-Purity Bacterial Cellulose

The transition toward a post-petrochemical economy has identified a critical bottleneck in the supply of raw biomass: the volatility of terrestrial agriculture. To stabilize the global material stream, we must pivot from the field to the fermenter. Microbial biomanufacturing—utilizing the intracellular and extracellular secretomes of bacteria, fungi, and actinomycetes—offers a path to synthesize high-performance polymers like bacterial cellulose (BC), exopolysaccharides (EPS), and gamma-poly-glutamic acid (gamma-PGA). Unlike plant-based materials, these polymers are uncoupled from climatic fluctuations and possess a molecular purity that precludes the need for toxic industrial bleaching.

1. The Purity Advantage: Bacterial Cellulose (BC) vs. Lignified Plant Tissue

Bacterial cellulose, specifically synthesized by strains such as *Acetobacter xylinum*, represents the absolute apex of polysaccharide purity. While plant-based cellulose is inextricably bound with lignin and hemicellulose—requiring aggressive chemical separation—BC is secreted as a pure, highly crystalline nanofibrillar network.

Deep Analysis: The Hydrogel Architecture

The structural superiority of BC is defined by its unprecedented water-holding capacity and high crystallinity. In medical engineering, this makes it an unparalleled matrix for 3D tissue engineering and artificial skin. Furthermore, BC possesses innate radical scavenging power, acting as a biological antioxidant. This allows the material to actively protect mammalian cells from reactive oxygen and nitrogen species (ROS/RNS), accelerating the kinetics of cellular regeneration and tissue repair.

2. Mycelial Architecture: From Brewery Waste to Automotive Composites

Fungal biomass, particularly the fibrous mycelium of macro-fungi and the spent biomass of commercial yeasts, represents a zero-cost industrial feedstock. The millions of tons of waste generated by the global brewing and fermentation industries can be chemically repurposed as a structural matrix.

- **Mycelium Composites:** By directing the growth of fungal mycelia through a substrate of agricultural waste, we can grow structural materials that replace toxic polyurethane (PUR) foams. These composites are lightweight, fire-resistant, and possess vibration-dampening properties.

- **Yeast Micro-Encapsulation:** Spent yeast cells are being utilized as biological micro-capsules for targeted drug delivery. Their cellular geometry allows for the protection of sensitive bioactive compounds and the delivery of nanoparticles directly to malignant tissues.

3. gamma-PGA and the Functionalization of the Secretome

The fermentation of glutamic acid by *Bacillus subtilis* yields gamma-poly-glutamic acid (gamma-PGA), a multi-functional, water-soluble biopolymer.

Deep Analysis: The Solubility Gradient

The gamma-PGA polymer is unique for its extreme hydrophilicity and biodegradability. In the environmental sector, it serves as a non-toxic flocculant for wastewater treatment, effectively replacing the synthetic polyacrylamides that currently contribute to microplastic accumulation. Its ability to sequester heavy metals via its carboxylate groups makes it a critical tool for the remediation of toxic industrial effluents.

4. Practical Advice for Bioprocess Engineers and Material Scientists

- **The Lignin-Free Mandate:** For medical-grade applications, prioritize BC over plant-based alternatives. The elimination of the bleaching stage reduces capital expenditure by over 25% and ensures the structural integrity of the fibers remains uncompromised.

- **Substrate Upcycling:** If operating a mycelium-based production line, establish a symbiotic pipeline with local breweries. Utilizing spent yeast as a secondary carbon source can reduce raw material costs to near-zero.

- **Acoustic Tuning:** When engineering BC for high-end acoustic diaphragms, strictly control the fermentation time. The thickness of the BC pellicle directly correlates with its response frequency; a 2–4 mm pellicle provides the optimal balance for high-fidelity audio reproduction.

The Arterial Waste-Stream: Phytological Sourcing and the Engineering of Lignocellulosic Composites

The transition to a post-petrochemical economy relies on the massive exploitation of the terrestrial biosphere. However, to achieve true sustainability, we must decouple material production from the global food supply. The arterial waste-stream represents the millions of tons of non-edible agricultural residues—sugarcane bagasse, wheat straw, and industrial sawdust—that currently remain underutilized. These residues are naturally rich in structural cellulose, hemicellulose, and rigid lignin, providing a nearly inexhaustible feedstock for advanced biopolymers.

1. The Phytological Matrix: Natural vs. Engineered Polymers

Plant-based materials are categorized into two distinct engineering streams based on their molecular origin:

- **Extracted Macromolecules:** These are polymers used in their near-native state. This includes hydrophobic zein, elastic gluten, and structural polysaccharides such as starch, agar, and alginate. Complex polyphenols like lignin provide the rigid scaffolding required for high-durability applications.

- **Synthesized Renewable Compounds:** These involve the fermentation of plant sugars into monomeric building blocks. Key examples include polylactic acid (PLA), polybutylene succinate (PBS), and the polyhydroxyalkanoate (PHA) family. Specifically, the copolymer PHBV has emerged as a premier drop-in replacement for polyethylene.

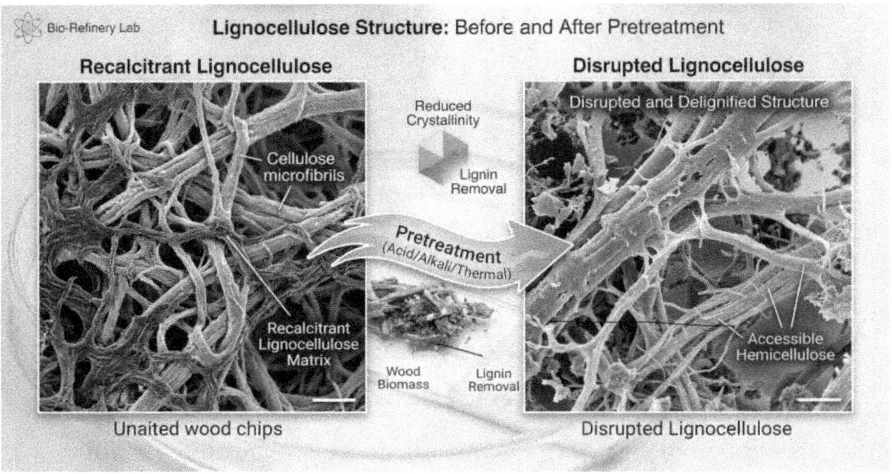

2. Biological Reinforcement: The Bamboo and Hemp Vanguard

To achieve industrial parity with synthetic materials, biopolymers must be reinforced with natural fibers that offer high tensile strength and low physical density.

Deep Analysis: The Density-to-Stiffness Ratio

Highly engineered natural fibers—such as jute, flax, hemp, and kenaf—provide a mechanical alternative to energy-intensive glass fibers. Bamboo fibers, with a physical density of approximately 1.4 g/cm3, fiercely compete with synthetic reinforcements in terms of specific stiffness. By utilizing aggressive mechanical milling, engineers can reduce macroscopic bamboo fillers down to nanoscale crystalline dimensions. These nanocrystals act as the ultimate biological rebar, providing bio-resins with the structural integrity required for automotive and aerospace applications.

3. Practical Advice for Bioprocess Engineers and Agronomists

- **Pre-Extraction Lignin Recovery:** Before processing straw or stalks for cellulose, implement a lignin-first biorefinery approach. Lignin is a high-value aromatic polymer that can be converted into bio-resins and adhesives.

- **Nanoscale Milling Optimization:** When developing bamboo-reinforced composites, utilize high-energy ball milling to achieve crystalline dimensions below 100 nm. This ensures a uniform distribution within the bio-polymer matrix.

- **The Non-Edible Mandate:** Prioritize feedstocks with high lignocellulosic complexity, such as cotton stalks. These materials offer superior mechanical properties for structural bioplastics and do not compete with the global caloric supply.

Comparative Material Performance Metrics

Feature	Petrochemical (PUR/PE)	Microbial (BC/Mycelium)	Phytological (Hemp/PHBV)
Origin	Fossil Fuel (Prehistoric)	Microbial Secretome (Grown)	Agricultural Waste (Arterial)
Purity	Refined (Bleached)	Absolute (Native)	Composite (Extracted)
Density	Variable	Ultra-Low	Low
Marine Half-Life	400+ Years	6 – 12 Months	6 – 12 Months
Primary Use	Insulation / Packaging	Wound Care / Speaker Cones	Vehicle Panels / Structural

The Commercial Vanguard: Market Penetration and the Geopolitical Scaling of Bio-Based Polymers

The global shift toward bio-based polymers has transitioned from a peripheral ecological niche to a central pillar of industrial strategy. Driven by the absolute sunset of single-use petrochemicals and a 600 million ton projected plastic demand by 2050, the biopolymer market is currently undergoing a period of hyper-aggressive commercialization. To succeed, these materials must not only mimic the physical performance of PE and PET but must also survive the forensic scrutiny of global lifecycle assessments (LCA) and strict legal definitions of decay.

1. Market Dynamics: The 33 Billion USD Threshold

The primary catalyst for this expansion is the European Union precedent. By instituting hard legislative bans on single-use plastics, the EU has forced the massive financial reorientation of chemical titans such as BASF, NatureWorks, and Braskem. This is not merely a market shift; it is a geopolitical pivot where regions that fail to adopt ISO-certified bio-based standards face total exclusion from high-value trans-Pacific and European markets.

2. The 'Greenwashing' Gap: ISO and ASTM Compliance

The most significant commercial risk in the biopolymer sector is the linguistic crisis surrounding biodegradability. As global trade authorities tighten regulations, the unqualified term "biodegradable" is becoming a legal liability.

- **The ASTM D6400 Mandate:** For a product to enter the industrial composting stream, it must be independently laboratory-tested against ASTM D6400 or ISO standards. These require the material to achieve total biological sanitization and fragmentation under managed, thermophilic conditions frequently exceeding 60°C.

- **The Home-Compost Paradox:** Unless explicitly certified for home-backyard composting, biopolymers like polylactic acid

(PLA) will persist in cold soil environments as physical pollutants. Companies that fail to differentiate between industrial compostable and home compostable face massive punitive fines for deceptive marketing.

The Jurisprudential Siege: Legislative Architectures for Global Marine Restoration

The saturation of the global marine environment with microplastics is not merely an environmental failure; it is a breakdown of international jurisprudence. As the oceans operate as the planet's primary regulating engine, the infiltration of synthetic polymers threatens the stability of all four essential ecosystem services: provisioning, regulating, supporting, and cultural. To resolve this, we must transition from a state of reactive litigation to one of structural interdiction, utilizing the principles of circular economics and extended producer responsibility (EPR) to sever the synthetic link between terrestrial industry and the marine biosphere.

1. The Architecture of Governance: From Linear to Circular

The fundamental objective of modern environmental law is the forced transition from a linear exhaust model to a regenerative circular economy. While legislation can effectively target macro-plastics through bans and collection mandates, the secondary microplastic fraction—created through the fragmentation of larger debris—is almost impossible to regulate once it enters the open ocean.

Deep Analysis: The Kinetic Regulatory Model

Effective governance must focus on the macro-micro nexus. By strictly regulating the production, use, and disposal of macro-plastics, we indirectly control the future baseline of microplastic accumulation. This requires a shift in focus toward extended producer responsibility (EPR), where manufacturers are legally and financially tethered to the entire life cycle of the polymer. When a corporation is responsible for the 100% reclamation or mineralization of its product, the market will naturally pivot toward bio-based, degradable alternatives.

2. The International Treaty Matrix: Specialized Interdiction

Global marine restoration is governed by a network of specialized conventions, each addressing a specific vector of infiltration:

- **MARPOL:** The primary barotropic boundary for maritime operations. It sets strict, legally binding targets to prevent the discharge of plastic waste from vessels.

- **CBD (Convention on Biological Diversity):** Under Article 14, the CBD mandates the reduction of plastic impacts on biodiversity through strategic environmental assessments.

- **The Basel Convention:** This allows for the regulation of the transboundary movement of plastic waste, preventing developed nations from exporting their synthetic pollution.

- **The Honolulu Strategy:** A collaborative framework between the UNEP and NOAA that represents a master blueprint for the prevention of marine debris.

- **The CMS (Bonn Convention):** This treaty identifies oceanic hot zones where microplastic density threatens migratory species and mandates national interventions.

3. Practical Advice for Policymakers and Environmental Engineers

To achieve a zero-pass standard for marine restoration, I recommend the following legal interventions:

- **The EPR Financial Mandate:** Implement a lifecycle bond. Manufacturers of non-degradable polymers must deposit a financial bond for every metric ton produced, released only upon verified recycling or mineralization.

- **Molecular Auditing of Wastewater:** Legally mandate the integration of MBR and RO membrane technologies for all municipal and industrial discharge points. Wastewater should be audited for polymer particle count.

- **The Arterial Law Strategy:** Focus legislation on river systems. Since rivers are the primary arteries feeding the ocean, national laws must treat riverine discharge with the same severity as direct ocean dumping.

The Regional Bastion: Chemical Jurisprudence and the Global Mesh of Polymer Regulation

The management of the global hydrosphere has shifted from broad, non-binding goals to a regional bastion model of high-resolution enforcement. As global treaties provide the skeletal framework, regional bodies—such as the UNEP Regional Seas Program and the EU Marine Strategy Framework Directive—provide the kinetic force required to neutralize synthetic flux. We are witnessing a paradigm shift in chemical jurisprudence: moving away from the assumption that polymers are inert due to their high molar mass, toward a forensic scrutiny of both the polymer matrix and the toxic additives embedded within.

1. The REACH Evolution: Beyond the 'No Data, No Market' Era

Historically, the REACH framework operated under a significant blind spot. Polymers were largely classified as polymers of low concern (PLC) based on the assumption that their high molar mass prevented biological uptake.

Deep Analysis: The Death of the PLC Myth

Modern toxicology has dismantled the PLC classification. We now recognize that while the polymer backbone may be stable, the associated additives—phthalates, flame retardants, and stabilizers—leach into the environment as the material fragments. Regional governance is now pivoting to a matrix-based scrutiny, where the registration of a polymer is contingent upon the toxicological safety of its entire chemical suite.

2. ECHA's Dimensional Siege: Redefining the Microplastic

The European Chemicals Agency (ECHA) has moved to establish a definitive dimensional siege against synthetic particulates. By redefining microplastics as any solid polymer particle where all dimensions are less than or equal to 5 mm, or synthetic fibers with a length of 3 nm to 15 mm, ECHA has effectively expanded the scope of industrial regulation to include the sub-micron and nano-scales.

The Regulatory Gap: The Invisible Flux

While current bans successfully target primary microplastics, a massive legal loophole remains. Secondary microplastics, specifically those generated from automotive tire wear and synthetic textile shedding, currently remain largely outside the reach of the ECHA dimensional standard. These sources represent the invisible flux that bypasses current wastewater infrastructure.

3. Practical Advice for Environmental Lawyers and Industrial Chemists

- **Additive Disclosure Protocols:** Manufacturers must move beyond proprietary trade secret labels for polymer additives. Pre-emptively disclosing the leaching kinetics of additives will be a prerequisite for market access.

- **Dimensional Labeling:** For products falling under the ECHA definition, implement dimensional forensic labels. These should specify the material type and the fragmentation potential under environmental weathering.

- **The Bio-Exemption Strategy:** Since truly biodegradable polymers are currently excluded from many ECHA restrictions, R&D funding should be redirected toward materials that mineralize within 12 months.

Comparative Analysis of Regulatory Frameworks

Framework	Scope	Primary Mechanism	Enforcement Level
MARPOL	Maritime / Vessels	Discharge Bans	International / Binding
Basel Convention	Transboundary Waste	Export Regulation	International / Restrictive
ECHA / REACH	Chemical / Product	Registration & Restriction	Regional / High-Resolution
EPR (Proposed)	Lifecycle / Producer	Financial Bonds	National / Economic

The Quantum Abyss: Colloidal Sovereignty and the Nanoscale Metamorphosis

The environmental trajectory of synthetic waste does not terminate at the microscopic level. We are currently witnessing a "Phase Shift" where microplastics, through the relentless violence of ultraviolet photo-oxidation and mechanical shearing, are being reduced to the sub-micron scale. These entities—scientifically designated as nanoplastics—represent a new, sovereign class of pollutants measuring strictly less than 1 µm. Unlike their larger progenitors, nanoplastics do not follow the predictable laws of Newtonian buoyancy. They have transitioned into the realm of colloidal physics, where they move with the chaotic autonomy of thermal energy and molecular collision.

The Physics of Invisibility: Brownian Motion and the Eco-Corona

In the open ocean, the distribution of macro-debris is dictated by density; it either floats or sinks. Nanoplastics, however, have effectively achieved "Hydrodynamic Sovereignty." Because their mass is so negligible, they are governed by Brownian Motion and electrostatic repulsion. They do not settle; they disperse with a perfect, haunting uniformity that saturates every cubic centimeter of the marine water column, from the sunlit surface to the crushing pressure of the Hadal zones.

A critical, yet poorly understood, phenomenon in this frontier is the formation of the Eco-Corona. Because nanoplastics possess immense surface energy, they do not remain "virgin" polymers for long. They aggressively engage in hetero-aggregation, fusing with natural organic matter and mineral ions to form complex, spongy architectures. This biological "disguise" allows the synthetic core to masquerade as a natural nutrient or a harmless colloid. This shell fundamentally alters the particle's toxicity, surface charge, and bioavailability, rendering traditional ecological models entirely obsolete. The nanoplastic is no longer just a piece of plastic; it is a hybrid, synthetic-organic entity that biological systems are not evolved to recognize.

The Biological Breach: Beyond the Blood-Brain Barrier

The most terrifying aspect of the nanoscale shift is the erasure of biological boundaries. While microplastics are largely confined to the gastrointestinal tract, nanoplastics are small enough to utilize cellular transport pathways. They possess the capacity to breach the blood-brain barrier, infiltrating the central nervous systems of marine species and inducing severe neurological deficits and behavioral pathologies.

We are moving from a crisis of ingestion to a crisis of infiltration. When a predator consumes a contaminated prey, it is not just taking on a load of plastic; it is absorbing a molecular-scale infiltrator that can embed itself within the muscular tissues, the liver, and the brain. The "Trojan Horse" has reached the cellular level, and the resulting toxicological "Black Box" is only beginning to be opened.

Practical Advice for the Next Generation of Researchers

The current reliance on physical filtration is an analytical dead-end for this size class. To bridge the current epistemological void, the scientific community must adopt a multi-modal analytical pivot:

- **Abandon the Mesh:** Stop attempting to use physical nets for sub-micron capture. Researchers should shift toward field-flow fractionation (FFF) coupled with multi-angle light scattering (MALS) to sort particles by their hydrodynamic radius rather than their physical size.

- **Trace-Level Annihilation:** Utilize pyrolysis-GC-MS optimized for trace-level detection to identify the polymer backbone of these sub-micron particles, as optical light (FTIR/Raman) often reaches its diffraction limit at this scale.

- **Histological Mapping:** Toxicologists must move beyond bulk concentration studies. We require in vivo histological profiling using fluorescently tagged nanoplastics to map the exact translocation pathways across cellular membranes and into the subcellular organelles (mitochondria and nuclei).

The Epistemological Void: The Black Box of Fragmentation

Despite the explosion of literature, we are operating in a state of fragmentation blindness. We do not yet understand how the original geometry of a plastic item—a flat film versus a dense sphere—dictates the terminal rate of its breakdown into nanoplastics. This geometric degradation rate is the missing variable in our global carbon and polymer models. Furthermore, the cross-boundary pathways—specifically how aerosolized nanoplastics move from the ocean spray into the high-altitude atmosphere—remain almost entirely unquantified.

Predictive Forecast: The Rise of Synthetic Viruses and Global Saturation

I predict that within the next two decades, we will identify a new class of synthetic pathogens. These will be nanoplastics that, through their specific size and eco-corona coating, perfectly mimic the physical and chemical markers of viruses. They will utilize viral entry receptors to gain access to human and animal cells, triggering localized apoptosis (cell death) and chronic inflammatory responses that current medicine is not equipped to treat.

Furthermore, I predict a global saturation milestone. As the inorganic smog of nanoplastics continues to rain down from the atmosphere onto agricultural lands, we will reach a point where the terrestrial food chain is as contaminated as the marine one. The loop will be closed not through the fish we eat, but through the vegetables grown in polymer-saturated soils and the very air we breathe. We are moving toward a future where the natural world is a geological memory, replaced by a unified, synthetic-biological hybrid metabolism. The ultimate terminal fate of plastic is not its disappearance, but its total, sub-micron integration into the fabric of life itself.

The Behavioral and Industrial Nexus: Consumer Sovereignty and the Permitting Fortress

The global landscape of plastic regulation is undergoing a fundamental inversion. We are moving away from a top-down model of bureaucratic mandate toward a bottom-up process driven by consumer sovereignty. This shift is occurring simultaneously with the construction of an industrial nexus—a permitting fortress that treats polymer production not as a commodity market, but as a high-risk chemical activity. To achieve the ambitious goal of 100% recyclability by 2030, the industrial metabolism must be aligned with the behavioral economics of the consumer-citizen.

1. The Consumer Sieve: Bottom-Up Policy Drivers

Modern plastic legislation is increasingly a reactive process, where policy follows the zenith of public awareness. As microplastic saturation becomes an inescapable reality in food wrappers, cigarette filters, and sanitary items, the consumer-citizen has emerged as the primary catalyst for absolute bans.

Deep Analysis: The Behavioral Economics of Incentives

Positive shifts in consumer behavior are rarely organic; they are the result of induced participation. Successful policy does not merely ask for reduction; it utilizes behavioral economics to make the alternative more convenient or financially beneficial. The role of the consumer is the material substrate of policy; without high-fidelity public participation, the most advanced EPR frameworks will suffer from a lack of collection velocity.

2. The Industrial Nexus: Permitting and Pellet Interdiction

For industrial enterprises, environmental permits have transitioned from administrative formalities to categorical mandates. In regions with high manufacturing footprints, the industrial nexus focuses on three critical pillars:

- **Hazardous Precursor Bans:** Strict prohibition of specific chemical additives and plasticizers that create permanent environmental liabilities.

- **Zero-Loss Pellets:** Legislation is increasingly targeting industrial pellet spills—the loss of raw resin beads during transport and processing. This represents a significant invisible flux that currently bypasses municipal waste oversight.

- **Informal Sector Integration:** Integrating shadow manufacturers into the formal waste-management permit regime is a non-negotiable requirement for regional marine restoration.

3. Practical Advice for Corporate Compliance and Consumer Advocacy

To navigate the transition toward the 2030 100% recyclability target, I recommend the following structural alignments:

- **The Digital Passport Audit:** Manufacturers should pre-emptively adopt digital plastic passports. By embedding a scannable QR or RFID tag into packaging, you can provide real-time transparency regarding recycled content and the environmental half-life of the material.

- **Pellet Containment Protocols:** For facility engineers, implement a closed-loop catchment system at every loading dock and extrusion point. Industrial pellet loss should be treated with the same severity as a chemical spill.

- **Incentive-Based Reclamation:** Policymakers should focus on gamified collection. By providing immediate digital micro-rewards to consumers for returning specific polymers, you can increase the reclamation rate far beyond standard curbside programs.

The Economic and Catalytic Lever: Extended Producer Responsibility and the CO2 Feedstock Revolution

The global failure to mitigate microplastic infiltration is primarily a failure of market internalities. In a linear industrial system, the ecological cost of a polymer is externalized to the commons. To achieve a zero-pass marine restoration, we must move beyond voluntary ethics toward a jurisprudential siege that utilizes economic instruments—specifically toxicity-based taxation and Extended Producer Responsibility (EPR)—and technological catalysts like CO2-based carbon-capture manufacturing.

1. The Economic Lever: Toxicity-Based Taxation and the Recyclability Index

Since the direct tracing of localized micro-emissions to individual firms is logistically complex, the most effective regulatory tool is the implementation of a molecular toxicity tax.

Deep Analysis: The Toxicity Hierarchy

Not all polymers are created equal in their environmental impact. Products composed of polystyrene (PS) and polyvinyl chloride (PVC) are empirically more hazardous to human health and ecological stability than other synthetics. A progressive tax framework should be levied based on a product's recyclability index:

- **High-Impact Polymers:** PS and PVC should face the highest tax rates to discourage their manufacture.

- **Sustainability Subsidies:** Industrial units that utilize bio-based feedstocks or demonstrate a 90%+ reclamation rate should receive direct governmental offsets.

2. EPR: Shifting the Molecular Liability

Extended Producer Responsibility (EPR) is a rights-based legal action that shifts the maintenance of the environment from the public sector back to the manufacturer. Under this framework, the producer retains molecular liability even after the initial sale.

- **Lifecycle Ownership:** Manufacturers are financially and logistically mandated to treat and dispose of their products post-consumption.

- **Market Professionalization:** EPR forces the growth of a professionalized, high-efficiency recycling sector. By assigning legal property rights to waste, the trash becomes an asset that the manufacturer is legally required to harvest.

3. The Catalytic Pivot: Carbon-Capture Manufacturing (PUR)

Polyurethane (PUR) is the world's third most widely utilized plastic, but its traditional production is carbon-intensive. A revolutionary shift in catalytic science now allows for the use of atmospheric CO_2 as a direct feedstock.

Analysis of the CO_2 Yield: For every 1 ton of epoxide replaced in the PUR chain, the process utilizes 3 tons of atmospheric CO_2. This carbon-capture manufacturing provides a threefold benefit:

- **Reduction in Baseline Footprint:** Significant decline in the environmental impact of raw material sourcing.

- **Active Sequestration:** The active removal of excess atmospheric carbon responsible for global heating.

- **Enhanced Durability:** PUR synthesized from CO_2 feedstocks shows superior resistance to chemical, thermal, and hydrolytic degradation.

4. Nano-Engineering and Biomimetic Barriers

One of the greatest challenges in recycling is the prevalence of non-recyclable multi-layered packaging (MLPs). Current research is focusing on nano-engineered, mono-material alternatives:

- **Magnetic Additives:** Utilizing magnetic particles to create air and moisture insulation, replacing the need for complex, bonded multi-materials.

- **Photosynthesis-Derived Cellulose:** Creating compostable, multi-layered barriers that mimic the functionality of aluminium-plastic foils but remain entirely biodegradable.

Comparative Analysis of Plastic Regulatory Instruments

Instrument	Mechanism	Target Phase	Economic Impact
Molecular Toxicity Tax	Progressive Taxation	Production	High (Cost-Push)
Digital Plastic Passport	Transparency / Blockchain	Usage / End-of-Life	Moderate (Admin)
EPR Mandate	Legal Lifecycle Liability	Post-Consumption	High (Investment)
CO2 Feedstock Subsidy	Direct R&D Offsets	Raw Material	Positive (Incentive)

The Final Synthesis: Strategic Outlook and Global Governance of the Plastisphere

The global oceans represent the most vital regulating engine of the terrestrial biosphere. Their health is not a localized environmental concern; it is a non-negotiable prerequisite for the survival of life. Currently, the plastisphere—the synthetic ecosystem created by the malignant spread of plastic pollution—threatens to permanently alter the nutrient cycles and biological productivity of the hydrosphere. To dismantle this synthetic siege, we must transition from fragmented national policies to a unified, global legislature that treats polymer persistence as a violation of international biosecurity.

1. The Governance Gap: Bridging the Developed-Developing Divide

A significant portion of global plastic leakage emerges from developing regions where waste management infrastructure is often non-existent or informal. The governance gap is currently the primary bottleneck in marine restoration.

Deep Analysis: The Failure of Geographic Isolationism

Plastic pollution is a trans-boundary fluid crisis; a failure of governance in a single river artery in Southeast Asia or South America has a catastrophic downstream effect on the entire global ocean. Strategic solutions must involve massive capacity-building mechanisms in the Global South. Extended producer responsibility (EPR) must be extended beyond metropolitan centers and down to the village level. This creates an inclusive circular economy where the manufacturer of the polymer is financially responsible for its recovery in every geography where it is sold.

2. The 4 Rs Re-Architected: From Individual Choice to Structural Mandate

The philosophy of the 4 Rs—reduce, reuse, replace, and recycle—must be redesigned as a structural reality of the industrial complex rather than an individual consumer choice.

- **Reduce & Reuse:** Mandating the elimination of low-utility polymers and the adoption of multi-use delivery systems.

- **Replace:** Forcing the market-wide transition toward the bio-based, 100% biodegradable alternatives detailed in previous sections.

- **Recycle:** Utilizing advanced molecular recycling and membrane technologies to ensure that the carbon-backbone of a polymer never fragments into the marine environment.

3. Comprehensive Executive Summary: The Jurisprudential Landscape

The paradigm shift in international law is anchored by sustainable development goal 14 (SDG 14) and the UN convention on the law of the sea (UNCLOS). This legal bedrock is supported by a specialized interdiction matrix:

- **REACH (EU):** Redefines microplastics as a chemical risk management issue, mandating the labeling of all solid polymers and additives. This ends the no data, no market era for synthetic macromolecules.

- **PWM Rules (India):** The 2021 amendments represent a landmark intervention, instituting a total ban on high-litter single-use plastics such as earbuds and expanded polystyrene.

- **Bottom-Up Pressure:** Modern legislation is increasingly a bottom-up process driven by consumer-citizens who utilize digital transparency to force a market-driven extinction of recalcitrant polymers.

The Behavioral Engine: Community Agency and the Deconstruction of the Plastisphere

Achieving a fundamental metamorphosis in human behavioral patterns is the absolute cornerstone of any successful strategy to dismantle the global microplastic crisis. Within the contemporary ecological framework, the human community occupies a paradoxical dual role: it is simultaneously the terminal victim of systemic toxicity and the ultimate point of origin for the 450 million metric tons of polymers produced annually.

1. The 5R Architecture: A Structural Scaffold for Remediation

The transition toward a circular economy is underpinned by the expanded 5Rs philosophy. This framework is not a set of lifestyle suggestions but a structural scaffold designed to bottleneck emissions:

- **Reduce:** Drastic curtailment of absolute plastic consumption, particularly in the single-use sector.

- **Reuse:** The implementation of infinite-loop packaging systems.

- **Replace:** The aggressive substitution of recalcitrant synthetics with bio-based alternatives.

- **Recycle:** The rigorous application of high-efficiency mechanical and chemical recycling protocols.

- **Recover:** The extraction of energy or high-value monomers from non-recyclable fractions.

2. The Hydrological Trap and the Collapse of Carrying Capacity

The most alarming dimension of the 21st-century plastic crisis is the role of the ocean as the hydrological trap. Because microplastic movement is a dynamic, fluid process, the pollution trapped in marine currents inevitably infests terrestrial biomes through atmospheric linkages and bio-accumulation.

Deep Analysis: The Inverse Law of Carrying Capacity

The carrying capacity (CC) of an ecosystem—its ability to fulfill basic human demands for clean water, food, and energy—exists in an inverse mathematical relationship with plastic saturation. Specifically, the carrying capacity is inversely proportional to the sum of macroscopic debris and microscopic particulates present within the environment.

As the density of macro-plastics and micro-plastics increases, the productive capacity of the environment plummets. This creates a lethal vicious cycle where human populations attempt to extract essential resources from a shrinking, deteriorated base, necessitating more synthetic inputs which in turn generate more waste.

The Feedback Loop of Resource Deterioration

This relationship highlights a critical threshold in environmental forensic engineering. When the accumulation of synthetic polymers reaches a critical density, the ecosystem's ability to provide essential provisioning services—such as uncontaminated protein from marine sources or clean groundwater—undergoes a non-linear collapse.

The resulting "Vicious Cycle" is an industrial-behavioral trap:

- **Resource Scarcity:** Declining natural yields force communities to rely on more intensive, synthetic-heavy production methods.

- **Secondary Infiltration:** These intensive methods, such as plastic-lined irrigation and synthetic fertilizers, introduce a secondary flux of polymers into the already stressed environment.

- **Carrying Capacity Erosion:** The cumulative synthetic load further depresses the natural recovery rate, leading to a permanent state of biospheric debt.

3. The Plastisphere: An Anthropogenic Fossil Record

We have catalyzed the creation of a global plastisphere—a novel, synthetic ecological layer extending from the Himalayan peaks to the Mariana Trench. This layer is fundamentally antagonistic to the natural biosphere. Every dimension of human development is currently being degraded by the infiltration of these particles. The plastisphere is essentially a non-native, anthropogenic fossil record being written in real-time across the planet's surface.

4. Practical Advice for Community Mobilization and Systems Engineering

- **The Elimination Mandate:** Shift local policy focus from beach cleanups to source interdiction. It is 10–20 times more cost-effective to stop a plastic bottle at the retail point than to recover its fragments from the benthic zone.

- **Hyper-Local IWMS Integration:** Deploy integrated waste management systems (IWMS) at the village and municipal level to reduce leakage velocity into local river arteries.

- **Citizen-Science Monitoring:** Equip community members with low-cost optical sensors and digital apps to track microplastic hot zones. This data provides the forensic evidence required to force local industrial units into compliance.

Global Sustainability Metric Comparison

Metric	Linear Economy (Status Quo)	Circular Bio-Economy (Proposed)
Material Origin	Prehistoric Carbon (Fossil)	Modern Atmospheric CO2
End-of-Life	Persistent Waste / Microplastics	Nutrient Return / Mineralization
Carrying Capacity	Declining (Inverse to Saturation)	Stabilized / Regenerative
Regulatory Model	Reactive Litigation	Structural Interdiction (EPR)
Economic Logic	Externalized Ecological Debt	Internalized Circular Value

The Demand-Supply Sabotage: Collective Agency as a Disruptive Force in Polymer Economics

The global plastic crisis is fundamentally a failure of the demand-supply feedback loop. The industrial production of recalcitrant polymers exists only because of a chronic, overwhelming public demand—a demand often manufactured through a convenience culture that externalizes the true ecological cost of the material. To disrupt this synthetic trajectory, the community must transition from being a passive consumer to an active causal agent of remediation. By strategically collapsing the demand for high-remanence plastics, collective agency can force an industrial pivot toward the bio-based architectures required for biospheric survival.

1. The Economic Architecture of Resistance

In a standard market model, production is a function of demand and the cost of raw materials. Because petrochemical feedstocks are heavily subsidized and the environmental end-of-life (EOL) costs are ignored, the current supply of plastic is artificially high. Grassroots movements act as a market inhibitor by introducing a new variable: the social license to operate. This variable increases the internalized cost of production. When the social license to operate drops toward zero, the industrial base is forced to either innovate or collapse.

2. Grassroots Jurisprudence: The Biocentric Model

The history of Indian grassroots movements provides a global gold standard for biocentric jurisprudence. These are not merely social protests; they are structural challenges to industrial homogeneity:

- **The Chipko Model:** Demonstrates the power of physical interdiction by directly placing the human body between the resource and its exploiter to preserve ecological integrity.

- **Beej Bachao Andolan:** A move toward biological sovereignty, protecting the genetic heritage of the planet from industrial monocultures.

- **Narmada Bachao Andolan:** A challenge to the infrastructure of displacement, proving that water security is a non-negotiable prerequisite for human welfare.

3. Three Dimensions of Direct Action

To effectively sabotage the current synthetic supply chain, collective agency must be channeled into three specific vectors:

- **Strategic Demand Suppression:** Consumers must target primary microplastics, such as microbeads and glitter, and non-recyclable multi-layered plastics (MLPs). By collapsing the market for these specific polymers, the industry is forced to adopt bio-derived alternatives.

- **Community-Led Remediation:** Moving beyond simple beach cleanups, communities must establish upstream interdiction points. By sorting and recovering polymers at the municipal level, we prevent macro-debris from entering the marine fragmentation cycle.

- **Political and Industrial Lobbying:** Concentrated pressure must be applied to enforce extended producer responsibility (EPR). When producers are legally mandated to own the molecular liability of their products, the economic incentive for producing recalcitrant waste evaporates.

4. Practical Advice for Community Leaders and Environmental Advocates

To maximize the impact of grassroots agency, I recommend the following direct action protocols:

- **Establish Data Sovereignty:** Do not wait for government reports. Utilize low-cost optical sensors and microscopy to document local contamination. When a community can prove exactly which industrial unit is responsible for a pellet spill, plausible deniability is stripped away.

- **The Pre-Cycle Mandate:** Shift community education from how to recycle to how to pre-cycle. This involves the total refusal of items with an environmental half-life exceeding 100 years.

- **Localized IWMS Scaffolding:** Create micro-sorter collectives. By professionalizing the informal waste sector and providing digital EPR credits for collected high-purity polymers, you turn a pollution problem into a regional economic asset.

5. Strategic Predictions: The Democratization of Forensic Evidence

I predict that by 2029, we will witness the global rise of **citizen-microscopy networks**. Smartphone-integrated spectroscopic tools will allow every citizen to act as a forensic environmental auditor. Communities will maintain live toxicity ledgers of their tap water and soil, linked directly to blockchain-based litigation platforms.

This democratization of data will spark a global wave of class-action lawsuits against the synthetic titans. The burden of proof will shift from the victim to the producer, forcing a market-driven extinction of high-persistence polymers long before legislative bans are fully ratified. The Plasticene ends when the consumer realizes they are the architect of the supply chain.

Comparative Impact of Collective Agency Strategies

Strategy	Primary Mechanism	Economic Effect	Environmental Outcome
Molecular Boycott	Demand Collapse	Reduced Production	Lower Micro-Emission
Physical Interdiction	Resource Blockade	Increased Capital Risk	Habitat Preservation

Strategy	Primary Mechanism	Economic Effect	Environmental Outcome
Data Sovereignty	Forensic Litigation	Liability Internalization	Industrial Compliance
Pre-Cycling	Material Refusal	Market Deselection	Zero-Waste Standard

The Cognitive Restoration: Pedagogical, Digital, and Grassroots Architectures for Synthetic Mitigation

The technical resolution of the microplastic crisis is impossible without a concurrent metamorphosis in human behavior. While the preceding sections provided the enzymatic, membrane-based, and legislative tools for remediation, this final architecture focuses on the behavioral engine. By mitigating the culture of excess through ecological literacy, grassroots intermediation, and the digital nexus, we can shift the global community from a state of passive consumption to active biospheric defense.

1. Mitigating the Culture of Excess: The 'Frugal with Grace' Philosophy

The systemic over-consumption of synthetic polymers is a sociological byproduct of the convenience era. To disrupt this, the community must adopt a multi-tiered strategy of selective refusal and material substitution.

Analysis of the Substitution Gradient: The transition away from polymers is not merely a rejection of plastic, but a re-adoption of materials with high molecular reciprocity. This includes high-grade borosilicate glass, 304/316 stainless steel, and natural protein fibers. Adopting a frugal with grace lifestyle prioritizes high-quality durability over disposable convenience, viewing the environmental half-life of a product as a primary functional value.

2. Pedagogical Frameworks: Translating Toxicological Data into Agency

A significant bottleneck in remediation is the information deficit. While scientific literature on polymer dynamics is abundant, it rarely reaches the community in an actionable format. Remediation fails when the public cannot differentiate between a polymer of low concern and a high-persistence toxicant.

Deep Analysis: The Ecological Literacy Gap

Pedagogical interventions must translate complex spectroscopic data and toxicological markers into community knowledge. This involves:

- **Legal Literacy:** Educating citizens on municipal codes and international ISO/ASTM standards, enabling them to identify and report industrial non-compliance.

- **Biological Literacy:** Understanding the link between microplastic infiltration (particles less than 5 mm) and the systemic disruption of the human endocrine and cardiovascular systems.

3. The Scientific Database: Risk Profiling and Molecular Fate

Rigorous research has established a clear gradation of risk for synthetic polymers. The terminal fate of microplastics and nanoplastics (particles less than 100 nm) is no longer a mystery; we have empirical proof of their ability to penetrate cellular membranes and disrupt reproductive health. Standardized methodologies for using bacteria and fungi to mineralize legacy waste are now moving from the lab to the community-scale micro-foundry.

4. Practical Advice for Community Leaders and Educators

- **Implement Plastic Footprint Audits:** Use digital toolkits to help households calculate their synthetic debt—the total volume of non-biodegradable waste they generate.

- **Localized Science Hubs:** Establish community microscopy labs via NGO partnerships. Allowing citizens to see the microplastics in their own local tap water is the single most powerful pedagogical tool for inducing behavioral change.

- **Legislative Advocacy:** Train community members to participate in municipal permit hearings to ensure no new industrial unit is granted a permit without a zero-microplastic discharge (ZMD) plan.

The Upstream Mandate: Industrial Metamorphosis and the Hazard Classification of Synthetic Polymers

The uninhibited proliferation of synthetic polymers represents a structural failure of industrial design. To effectively mitigate the microplastic crisis, we must pivot from end-of-life management to upstream interdiction. This industrial metamorphosis requires a fundamental re-architecture of how materials are conceived and manufactured, ensuring that the cost of persistence is internalized at the point of production.

1. The Five Pillars of Industrial Structural Intervention

The transition toward a biospheric-compatible industrial base relies on five distinct strategic pillars:

- **Material Substitution:** Aggressively adopting glass and bio-sourced materials in sectors with marginal functional utility for plastics.

- **Architectural Design Optimization:** Utilizing generative AI to decrease the absolute mass of plastics used in each product unit without compromising structural integrity.

- **Longevity and Repairability:** Engineering products for an increased usability age by facilitating modular replacement.

- **Elimination of High-Frequency Pollutants:** Implementing a total industrial sunset for high-risk single-use plastics (SUPs).

- **Recyclability Engineering:** Strictly limiting the number of different polymers and toxic additives used in a single product.

2. The Economic Disparity: Internalizing the EOL Deficit

A significant economic hurdle remains: recycled products are frequently more expensive than virgin petrochemical feedstocks. This is a market distortion caused by the failure to price the environmental half-life of the material. By internalizing the end-of-life (EOL) costs through mandatory recycling fees, the price of virgin petrochemicals will naturally rise to meet the cost of sustainable alternatives, creating true market parity.

3. Hazard Classification: The Sunset of Toxic Polymers

The current regulatory framework treats most plastics as inert until they fragment. This is a catastrophic oversight. Certain polymers possess inherent chemical toxicities that mandate a high-level reclassification.

Hazardous Candidates: I anticipate that within the next decade, polycarbonate (PC), polyvinyl chloride (PVC), and polystyrene (PS) will be globally classified as hazardous substances. This classification will trigger absolute bans based on their impact on biospheric health through the leaching of phthalates and bisphenols.

4. Practical Advice for Corporate Strategy and EPR Implementation

- **Transparent Additive Labeling:** Implement a molecular ingredient list for all polymer products. Identifying every stabilizer and flame retardant used is essential for determining the toxicity profile.

- **The Recycling Fee Integration:** Manufacturers should pre-emptively integrate a unit-based recycling fee into their pricing models to fund localized recycling infrastructure.

- **Mono-Material Optimization:** Eliminate the use of complex mixers and multi-layered plastics (MLPs). A product that cannot be mechanically or enzymatically separated should be viewed as a design failure.

Strategy	Primary Mechanism	Target Objective	Impact Scale
Cognitive Restoration	Pedagogical Literacy	Behavioral Metamorphosis	Community-Wide
Upstream Interdiction	Material Substitution	Demand Suppression	Industrial-Level
Hazard Reclassification	Legislative Mandate	Toxicity Neutralization	Global / Regional
Digital Accountability	Transparency Engine	Market Deselection	Consumer-Driven

The Structural Interception: Thermodynamics, Chemical Recycling, and the Mirage of Circularity

The transition from a linear take-make-waste model to a circular economy is currently hindered by the thermodynamic reality of polymer degradation. To mitigate the marine saturation of microplastics, we must view the journey of synthetic waste not as a disposal problem, but as a molecular journey requiring constant structural intervention. While integrated waste management systems (IWMS) provide the physical scaffolding for recovery, the ultimate success of remediation depends on our ability to overcome the entropy of mechanical recycling and the economic friction of virgin petrochemical subsidies.

1. Structural Interception: The IWMS Architecture

The journey of plastic waste—from industrial synthesis to environmental leakage—is a pathway of increasing entropy. To intercept this journey, we must implement a three-tiered integrated waste management system (IWMS):

- **High-Quality Chemical Recycling:** The primary objective is to break polymers back down into their original monomers via solvolysis or pyrolysis, bypassing the physical limitations of mechanical heat cycles.

- **Feedstock Energy Recovery:** For non-recyclable fractions, high-temperature incineration with flue-gas scrubbing serves as a secondary option, converting the carbon-backbone into thermal energy.

- **Inert Mineralization:** The final terminal fate of truly unusable fractions must be restricted to mineralization ash to ensure no non-biodegradable polymers enter the water column.

2. The Circularity Mirage: Mechanical vs. Chemical Realities

Recycling is often marketed as a flawless loop, but it is in fact a complex, energy-intensive process plagued by polymer incompatibility. When different polymer types are cross-contaminated during the melting process, their chemical signatures conflict, significantly compromising the mechanical integrity of the resulting product.

Analysis of the Recycling Gradient:

Primary recycling yields virgin-equivalent plastics from uncontaminated industrial scrap. However, most consumer recycling falls into the category of secondary recycling or downcycling. This involves transforming contaminated waste into lower-grade materials like plastic-reinforced asphalt. This is not true circularity; it is merely a waste delay where the polymer eventually fragments into microplastics at its secondary end-of-life.

3. Thermodynamic Optimism: The Energy-Efficiency Gap

Despite the hurdles, the physics of recycling offers a significant energy dividend. For polymers such as PET and PE, the energy required for mechanical recycling is approximately 50% lower than the energy needed to synthesize virgin resins from crude oil.

Deep Analysis: The Five Bottlenecks of Circularity

The slow pace of global circularity is driven by five identifiable thermodynamic and economic frictions:

- **Thermal Scission:** Every heat cycle in mechanical recycling causes the polymer chains to shorten, degrading tensile strength.

- **Subsidized Feedstocks:** The low price of crude oil makes virgin resin more financially attractive than recycled pellets.

- **Chemical Cross-Contamination:** The presence of PVC or PS in a PET stream can render the entire batch industrially useless.

- **Cycle Limits:** Most polymers can only be mechanically recycled 3 to 5 times before their molecular integrity collapses.

- **Hybrid Obstruction:** The proliferation of multi-layered laminated films creates a separation barrier that current facilities cannot breach.

The Synthetic Recession: A Multi-Generational Roadmap for Biospheric Restoration

The accumulation of microplastics in the global hydrosphere is the definitive systemic illness of the industrial era. However, the plastisphere is not a permanent geological feature; it is a temporary historical anomaly that can be dismantled through the synergy of community agency, industrial re-architecture, and bioengineered remediation. To achieve a synthetic recession by the mid-21st century, we must execute a multi-generational strategy that shifts the global industrial complex from a petrochemical-based linear economy to a bio-integrated circular one.

1. The Temporal Roadmap: Milestones for Restoration

The reversal of synthetic saturation requires a phased approach, categorized by the following kinetic milestones:

- **Short-Term Objectives (0–5 Years):** Immediate cessation of high-frequency single-use plastic (SUP) production and the mandatory replacement of microbeads with biodegradable cellulose or starch-based alternatives.

- **Mid-Term Objectives (5–15 Years):** Global implementation of IWMS and the universal adoption of extended producer responsibility (EPR).

- **Long-Term Objectives (15+ Years):** Transitioning the manufacturing sector entirely to renewable energy and CO_2-captured feedstocks.

2. The Definitive 5R Philosophy: A Structural Mandate

The future of human civilization depends on a lifestyle and industrial architecture anchored by the 5Rs. This is no longer a personal choice but a structural requirement for biospheric survival:

- **Reduce:** Shrink the absolute mass of polymers produced per capita.

- **Reuse:** Extend the functional usability age of every polymer unit through modular repairability.

- **Replace:** Transition aggressively to certified compostable and bio-based materials like PHA, PHBV, and PLA.

- **Recycle:** Ensure 100% of synthetic carbon is recovered via high-efficiency solvolysis or pyrolysis.

- **Recover:** Extract energy and valuable precursor chemicals from the terminal waste stream.

3. Practical Advice for Industrial and Community Leaders

To navigate the way ahead, I recommend the following strategic imperatives:

- **The Zero-Landfill Industrial Audit:** Manufacturing firms must transition from voluntary measures to mandatory accountability. Implement a unit-based recycling fee for every unit of plastic sold to fund a professionalized recycling sector.

- **Localized IWMS Scaffolding:** Communities should prioritize hyper-localized waste management. Preventing a plastic bottle from fragmenting in a local river is 1,000 times more efficient than attempting to filter nanoplastics from the open ocean.

- **The Hazard Reclassification:** Lobby for the formal global classification of PVC, PS, and PC as hazardous substances to trigger the market-driven extinction of these high-persistence polymers.

Comparative Efficiency: Linear vs. Molecular Circularity

Metric	Linear Economy	Mechanical Recycling	Molecular Harvesting (Solvolysis)
Energy Input	100% (Baseline)	50% Reduction	70% Reduction (Theoretical)
Material Integrity	Virgin Grade	Degraded (Thermal Scission)	Virgin Grade (Monomer Recovery)
Cycle Limit	Single Use	3 – 5 Cycles	Infinite
Contamination Tolerance	Zero	Low	High (Chemical Separation)
Marine Impact	High	Delayed Fragmentation	Zero (Total Erasure)

The Forensic Sieve: Retrospective Remediation and the Engineering of Autonomous Ocean Recovery

The saturation of the global hydrosphere with synthetic debris necessitates a dual-track technological response: prevention and collection. While prevention is thermodynamically and economically superior, the hundreds of millions of metric tons already circulating require aggressive retrospective remediation. Crucially, the extraction of macro-plastics serves as the primary defense against microplastic proliferation; every 1.5 kg bottle removed today prevents the formation of billions of future micro-shards.

1. The 'Boats and Wheels' Paradigm: Energy vs. Entropy

There is a distinct architectural split in the global response to plastic pollution. Prevention technologies are largely stationary filters embedded in infrastructure, whereas collection technologies rely on boats and wheels—mobile, energy-intensive units such as aquatic drones and trash skimmers.

Deep Analysis: The Efficiency Gradient

The current technological landscape is heavily skewed toward macro-scale debris. This is driven by the physics of buoyancy and concentration. It is orders of magnitude more energy-efficient to capture a floating 1.5 kg container than it is to filter millions of liters of water to extract a matching mass of neutrally buoyant micro-shards. However, the reliance on mobile vessels creates a high carbon cost for remediation, necessitating a shift toward passive or autonomous energy-neutral systems.

2. Forensic Attribution: Solving the 'Unrecognizable' Residual Load

Effective remediation depends on the ability to categorize particles into primary microbeads and secondary fragmented debris. When particles exhibit distinctive traits—such as the perfect sphericity of beads or the fibrous geometry of textiles—the industrial vector of emission can be pinpointed. However, a significant proportion of environmental microplastics are so highly weathered that they are unrecognizable. This residual load requires the integration of advanced spectroscopic sensors to identify chemical signatures even when physical morphology has collapsed.

Index

A

- **Acoustic Microfluidics**
- **Additives (BPA, phthalates, flame retardants)**
- **AFM-IR (Atomic Force Microscopy-Infrared)**
- **Air purification / textile smog**
- **Albedo shift (oceanic)**
- **Anaerobic metabolism / methane**
- **Antibiotic resistance / plastisphere superbugs**
- **Aquaporin biomimetic membranes**
- **Arterial interdiction (river / wastewater gates)**

B

- **Bacterial cellulose (BC)**
- **Benthic graveyard / deep-sea sink**
- **Biodeterioration / surface erosion**
- **Bio-encrustation / biofouling**
- **Biological Trojan / synergistic toxicity**
- **Biomonitoring / histological forensics**

- Bioplastics / biogenic polymers

- BPA (Bisphenol-A) & phthalates

- Brine disposal / plasma gasification

- Buoyancy paradox / bio-anchoring

C

- Carbon sink sabotage / photonic starvation

- Cascade of exclusion (mechanical sieving)

- Chemical coalescence (organosilane fixatives)

- Clean-room mandate / QA/QC protocols

- Clinical screening / body burden analysis

- CO_2 feedstock / carbon-capture PUR

- Compostability (ASTM D6400 vs. home)

- Crystallinity barrier / amorphous zones

- Cytometry (machine-learning enhanced)

D

- Density-based flotation / heavy brines

- Dynamic membrane (DM) / cake-layer filtration

E

- ECHA / REACH microplastic definition

- Eco-corona / colloidal sovereignty

- Encrustation (synthetic)

- Endocrine sabotage / genomic weathering

- Enzymatic cocktail / exo- vs. endo-attack

- Enzymatic insurgency / molecular guillotine

- EPR (Extended Producer Responsibility)

F

- Field-flow fractionation
- Forensic identification / Pyr-GC-MS
- Forward osmosis
- Fragmentation cascade
- Froth flotation
- Fungal bio-foundry / mycelial siege

G

- Genomic weathering / transgenerational EDCs
- Gravimetric defiance / buoyancy threshold

H

- Heavy brines (zinc chloride, sodium iodide)
- Horizontal gene transfer (HGT)
- Hybrid biosphere / plastisphere taxonomy
- Hydrophobic material design / low-sorption polymers
- Hypoxic micro-zones / metabolic tax

I

- In-situ autonomous spectrometers
- ISO-certified clean-room protocols

K

- Kinetic ceiling / stoichiometric barriers

L

- Lab-on-a-Chip / pinched flow fractionation

- Lethal heat spikes / thermal blanket

M

- Machine-Learning Enhanced Cytometry
- Manta trawl / flowmeter
- Membrane bioreactors (MBR)
- Metabolic suffocation / synthetic bio-colony
- Microplastics (large / small)
- Molecular incineration / thermal dissection
- Molecular trespass lawsuits
- Mycelial infiltration / hyphal advantage

N

- Nanoplastics / nanoscale soup
- Nurdles (primary resin pellets)

O

- Oxidative stress / cellular damage

P

- PHA (polyhydroxyalkanoates)
- Photonic starvation / artificial canopy
- Plastisphere
- Polymer terminal / molecular colonization
- Primary vs. secondary particles
- Pyr-GC-MS & additive profile

R

- Reverse osmosis (RO)

S

- Saponification (KOH) / H_2O_2 oxidation
- Secondary particles / fragmentation
- Smart porosity interface
- Sterile bastion / synthetic enclosure
- Synthetic bio-colony
- Synthetic canopy / thermodynamic disruption
- Synthetic encrustation
- Synthetic gastronomy / dietary vector
- Synthetic metabolism / recombinant consortia
- Synthetic rain / atmospheric polymerization
- Synthetic sieve / membrane geometries
- Synthetic trophic shift

T

- Thermal blanket / convective loss
- TGA-FTIR-GC-MS
- Trophic transfer / bio-accumulation

U

- Ultrafiltration (UF)

Z

- Zero-sorption mandate